T0179946

# A.W. (BUCK) SAUNDERS

## LIEUTENANT COLONEL U.S. AIR FORCE RETIRED

# THE PRICE

# OF

# GLORY

### A MILITARY AUTOBIOGRAPH

1941 – 1965

### THREE WAR'S

SECOND WORLD WAR

KOREAN WAR

VIETNAM WAR

Turner Publishing Company

ISBN: 978-1-56311-801-2
Library of Congress Control Number: Applied For
2002103209
Project Coordinator/Designer: Herbert C. Banks II.
Copy Editor: Hallie Belt

Limited Edition.

# MY CREDO

*There Is an Appointed Time for Everything;*

*A Time to Be Born and a Time to Die...*

*A Time to Love and a Time to Hate;*

*A Time of War and a Time of Peace.*

– Ecclesiastes 3 : 1-8

# TABLE OF CONTENTS

# PHOTOGRAPHS

# ON SILVER WINGS

Oh silver wings, please carry me home.
On silver wings, I'll never more roam.
I've wandered far, o'er land and o'er foam
On silver wings, that taught me to roam.

I've seen the roads, to Mandalay
The harbor lights, of old Bombay
Oh silver wings, I'm so all alone
So please make haste, and carry me home.

I met a gal, down Tennessee way
and promised her, I never would stray.
But the cannons roared, and I knew I must go
on silver wings, to meet with the foe.

If I should die, before it ends
I pray the Lord, forgive my sins.
and to my loved ones, let it be known
that silver wings, have carried me home.

# FOREWORD

*The Price of Glory* is an autobiography of my twenty-three-plus years in the Army Air Corps and the Air Force. I have tried to show the good, the bad, and the ugly sides of military life as an enlisted man and as an officer.

I have tried to point out the abject stupidity of some of the senior officers and the decisions they made, as well as the incompetence of many of the unit commanders and their inability to lead. In my opinion, too many assignments were made based on the rank of the individual instead of his ability to do the job.

It has been said that flying is hours of boredom interspersed with periods of excitement and moments of stark terror. I have experienced it all.

What is surprising to me is that we ever won a war. I don't think we did. I think that the enemy just blew it.

# Chapter I

## REINCARNATION

I must have been reincarnated from a warrior of long ago because I spent much of my childhood playing war. Born in Sonora, Texas, I lived there until I was nine. My cousin, who lived on a ranch about 7 miles away, was a lot like me. He liked to play war games, too. He probably inherited his love for the military from his father. During the first war, his father wanted to join the army and go to war, but he was told he was too underweight, so he ate bananas and drank cream to increase his weight so the army would take him and he could go fight. When he came back from Germany, he brought a German helmet and bayonet. My cousin and I used these in our war games. We had so many trenches and fox holes dug in his backyard, it wasn't safe to go out there at night without a light.

In the late 1930s, the U.S. military had what they called "Civilian Military Training Camps" (CMTC). They were for thirty days each summer. After four summers, you were given a second lieutenant's commission in the Army Reserve. After graduating from high school, my cousin attended three of the camps. In 1940, he begged me to go to camp with him at Camp Bullis, near San Antonio. It didn't take much arm-twisting because of my love for the military. I thought that a soldier's life would be the most glorious life a man could have, but I didn't know *the price of glory*.

In 1940 I signed up for the Camp at Camp Bullis and was assigned to the 12th Field Artillery unit there. My cousin was in an infantry unit.

One day after lunch, my cousin and I were resting in our tents before going back for more training. I started having bad stomach cramps, so I went to the first sergeant, an old thirty-year man, and told him of my problem. He said he would take me to the dispensary to see a doctor, but I would have to wait in line, as they were checking men for sore throats.

When I finally got to the doctor, he checked me over, gave me some medicine, and told me to stay in my tent. I went back to my tent, took some of the medicine, and lay on my bunk. By five o'clock, I hurt so much I couldn't stand it, so I went back to the first sergeant, and he took me back to the dispensary. I had to wait in line again. When I got to the doctor, he shook down a thermometer, put it in my mouth, and went through a side door. He came back forty-five minutes later (he had been to chow), so the old sergeant gave him a good cussing. The lieutenant said, "Sergeant, I'll have you know you are talking to an officer."

The sergeant said, "I may be, but you're a sorry example of an officer or doctor. Anyone who would treat a man as sick as this one, the way you have, should not be a doctor or an officer."

This comment made the lieutenant mad, and he told me, pointing to the examining table, "Get up on this table." I crawled up on the table and lay on my back. He hit my stomach so hard with the fingers of both hands, I rolled off the table. He said, "Acute Appendicitis. Sit down over there. Next." I sat for a while but couldn't sit there any longer. I got up and started walking back and forth, all doubled over. In a loud voice, the lieutenant said, "Orderly." When the orderly came in, he said, "Show this man a bed. He's getting on my nerves." In the back of the ward, there were about twenty beds. The orderly took me back there and told me to get on one of the beds. There was no one else there. No nurse, no orderly, no patients, no one.

I didn't see another person until about eleven o'clock that night, when the side doors opened and an ambulance backed up to the door. An orderly came in and said, "Come on. We're going to Brooke Hospital at Fort Sam." I couldn't sit up, so I rolled off the bed onto the floor. I got up on my knees and was able to stand up and get to the ambulance. I crawled on a stretcher, they closed the doors, and we left.

When the ambulance got to Brooke Army Hospital at Fort Sam Houston, it backed up to the emergency entrance. Two men carried me in and put me on the examining table.A major came over, gave me a quick examination, and said, "Rush this man to surgery." They took me up to the fourth floor, where a nurse gave me a shot to ease my pain, prepared me, and rushed me to surgery for an emergency appendectomy.

The next morning, the major who had examined me and performed the operation was making his rounds. He came over to my bed, checked my chart, and asked, "How do you feel?" I told him I felt a lot better than I had the evening before. He said, "You know, you almost didn't make it. Your appendix had ruptured."

I said, "I knew I was awfully sick." I then told the major what had happened at the dispensary at Bullis: I had not seen a single person from about 6:30 until about 11:00 p.m., and I could have died there. "Major," I said, "when I get out of this hospital, I'm going to kill that lieutenant. No one can treat me like that, risk my life like that, and get away with it."

They kept me on the ward for seven days and then sent me over to the old hospital for fourteen days of recuperation. (At that time, they kept appendectomy patients for twenty-one days.)

When I was released, they took me back to Camp Bullis. I went to my tent, strapped on a 45 pistol, and returned to the dispensary to look for the lieutenant, but he was gone. I never knew if the major had him booted out or what had happened to him. Whatever it was, it kept me from making a big mistake. I think I would have shot him if I could have found him, and I would never have had a career in the military. At the end of the training, I went back to Sonora.

In the spring of 1941, I again applied for CMTC training. I was told the program had been suspended indefinitely.

# Chapter II

## ACTIVE DUTY

War clouds were gathering on the horizon. The situation in Europe was going from bad to worse. One day I told my mother, "I think the U.S. will be at war before the end of the year, and I want to be in it. I want to fight for my country." She knew how I felt about the military, so she didn't try very hard to talk me out of joining the Army Air Corps.

On May 20, 1941, I went to San Angelo, Texas, and enlisted in the Army Air Corps. They sent me to Fort Sam for processing. I was then sent to Lowry Field, Colorado, for boot training and waiting assignment. It was the first time I had been out of Texas, except for going across the border to Old Mexico.

They were just building Lowry Field, so we had to live in tents. They made me an "Active Drill Sergeant" because of my training at Bullis. The other recruits had had no training at all. We drilled all month so we could put on a good show for the VIPs from Washington. We would have several thousand men marching on the ramp in the parade. After the parade, the planes would come over and put on an air show. It was very impressive, and it made me want to be a pilot. Little did I realize at the time that I would wind up as a pilot in the Army Air Corps and, later, the Air Force for twenty-three years. At the time, I was a private, getting $21 a month.

When you go into the service, they interview you to see what your training and background is so they can determine what field of work would be best for you. I had worked on repairing engines and was mechanically-inclined. Because of that, they sent me to the Airplane Mechanic School at Chanute Field, Illinois. It was a five-and-a-half-month school, the best in the country. I considered myself lucky to get the assignment.

While at Chanute, I saw the northern lights for the first time. They were the most beautiful sight I had ever seen. At night, they looked like giant veils or curtains of colored silk hanging in the sky. Then they would seem to race across the sky in every direction at once. Their beauty begs description. I wish everyone could see the northern lights.

I was thoroughly enjoying the school. However, I decided that if I was going to make a career of the military, I might as well be an officer. If I were an officer, maybe some day I could go to flying school and become a pilot. I had wanted to fly ever since I was a young boy. When I was about nine years old, I built an airplane out of scrap lumber that I could sit in and pretend I was flying.

I decided to apply for Officer Candidate School (OCS). After ninety days there, one can become a second lieutenant or "90-Day Wonder." To apply, you had to submit letters of recommendation. I wrote to my grandfather to see if he could get me some high-powered letters of recommendation from men he knew in Austin. He was well-known there, being the most senior land surveyor in

Texas, so he sent me letters from the general land commissioner and the state attorney general. I knew it wouldn't hurt to use a little political pull in applying for the assignment. However, before I had all of the paperwork ready to submit the application for OCS, I heard about a flying sergeant program they had just started. They were accepting enlisted men who had a high school diploma and were sending them to flying school, where they would graduate as staff sergeant pilots. If you had at least two years of college, you could apply for the flying cadet program and graduate as a second lieutenant. I didn't have the college, and since being a pilot was my ultimate goal, I used the letters of recommendation to apply for the flying sergeant program.

In December of 1941, I graduated from the Airplane Mechanics School and was assigned to Bradley Field at Windsor Locks, Connecticut. Bradley Field had the distinction of being the only "camouflaged field" in the U.S. All the taxiways and the runway were painted with a camouflaged design, and the buildings were built like farm buildings.

I was assigned as a crew chief on a photo reconnaissance plane. There was only one hangar on the field, and it was for a P-40 fighter unit. We didn't have a hangar in which to work on our planes, so they were kept in sandbag revetments. All of the work on the planes had to be done out in the open. There was so much snow that the walkway from the barracks to the work area was like an open-top tunnel. The temperature dropped as low as -20° F, and my hands and feet got so cold, they had no feeling. I remember one day one of the men accidentally stepped on my hand, and I didn't even feel it.

In January 1942, I received the long-awaited word that I had been accepted for pilot training. The pilot of the plane I was crewing started letting me fly it after we were airborne, and he would show me how to do various maneuvers. One day he said to me, "I'll give you some advice that was given to me when I started flying that I found to be excellent advice: Know your limitations and the plane's limitations, and stay within both, and you will always come back." In my twenty-three years of flying, I found that in most aircraft accidents, one or both of those limitations had been exceeded. It was the best advice I have ever received.

# Chapter III

## SECOND-CLASS CITIZEN

In February 1942, I was sent to Maxwell Field Alabama for preflight training. We were called aviation students, as opposed to aviation cadets. We were billeted in an abandoned old mill building near the base. The cadets in the flying program were billeted in new barracks on the base. Whereas the cadets lived in rooms, we lived in an old warehouse-like building, slept on cots, and hung our clothes on a wire stretched across the building. The Old Mill was an open building with no partition walls, except for the latrines. The canvas cots were close together, and stretched wire was beside each row of cots for us to hang our uniforms on. There was room at the foot of each cot for a footlocker, where we kept our socks, underwear, toilet articles, etc.

It was unbelievable that we would be subjected to such conditions. I should have gotten a clue from this that we were to be second-class citizens in the Air Corps, even though we were to take the same training and courses as the cadets *and* fly the same planes. We got into flying because we loved to fly, whereas the cadets for the most part got into flying to avoid the draft. The cadets were to graduate as officers, and we were to graduate as enlisted pilots. It was a stupid concept but was only one of many stupid concepts I would encounter in the years to come.

Although we had to live in the Old Mill while the cadets lived in new barracks, we didn't mind too much. We could look down the road to the day when we would be pilots. We could put up with anything, as long as we could be pilots. At least the food was good, and the place was a lot warmer than Bradley Field.

In March of 1942, I completed preflight training and was sent to a primary flying school at Madison, Mississippi. We were to train in the PT-17 Primary Trainer. I made my first flight on March 30. Flying came very easy for me, partly because the captain at Bradley had given me the instruction in the F-2. As my instructor said, I seemed to have a natural ability for flying. He cleared me for solo after I had had only eight hours of instruction.

While I was at Madison, my oldest brother came to see me. He had also enlisted in the Air Corps and was taking the airplane mechanics course at Kessler Field, Mississippi. It was the same course I had taken at Chanute Field. When he went back, I went with him into Jackson to see him off. Who should we run into but the commandant of students at Madison. I had failed to get permission to leave the base, and he asked me my name. When I got back to the base, I was given ten hours of "tours" marching on the ramp with a parachute on my back as punishment for being off the base without permission.

About halfway through the program, there was a rumor that about half the class would be sent to glider training. I had heard that about half of the glider pilot students were killed during training, and all of the students started talking about it. No one wanted to go to glider training. We decided it would be better to wash out of pilot training than go to glider training. We thought if we buzzed Jackson or did acrobatics over the city, we would be washed out and sent back to our old base. That would be the end of our dream, but at least we wouldn't be killed in a glider. Somehow the commandant found out what we were thinking about doing, and he stopped all solo flying. We could only fly with our instructors. After a few days, he realized that he had to let us keep training, so he put out the word that no one would be sent to glider training.

While I was at Madison, another incident occurred involving the Cadet Club that was in the Heidelberg Hotel in Jackson, about 10 miles from Madison. Back in 1942, the military still permitted the upper class to haze the lower class. Because of the way they treated us, there was no love lost between the two classes. Because of the animosity between the two classes, they were never permitted to use the club at the same time.

One day, the commandant decided to let the two classes use the club at the same time, an act destined for failure from the beginning. That weekend, when the club opened, the brawl that ensued between the two classes resulted in the breaking of a lot of the furniture and complete disruption of the decorum of the hotel. The next day, the commandant called a meeting of all the students. He told us, because of our actions, we owed the good people of Jackson a form of apology. He decided that an appropriate gesture on our part would be for us to march from Madison to Jackson and put on a parade.

The day before we were supposed to march, he announced that the parade had been called off. I don't think he had ever intended to go through with it. He just wanted to scare us and teach us a lesson in how aviation students should conduct themselves. He never again let both classes use the club at the same time.

# Chapter IV

## A DREAM COME TRUE

In May of 1942, I had completed Primary Training. Many of my classmates who just couldn't hack it had washed out of the course. I was sent to Greenville, Mississippi, for Basic Training. We trained in the Vultee BT-13 plane. ("BT" stood for "Basic Trainer," and it was called the "Vultee Vibrator" because it shook so much on the ground.) I made my first flight in the BT-13 on June 4, 1942. By this time, I had felt completely at home flying.

One day, about half the class came down with food poisoning. Fortunately, I was not affected. Several ambulances were shuttling between the barracks and the hospital, taking those who were sick. Other than that, Basic went smoothly.

I was in Greenville one day when a truck went by with what was left of a Crop Duster plane. The tail was the only thing that looked like a part of a plane. I found out that a sixteen-year-old boy was flying it when it crashed—and walked away without a scratch.

In August, I completed Basic Training and was sent to Columbus Air Force Base for Advance Training, which was the final phase. I knew that if I finished Advance, I would finally be a pilot, and my dream would have finally come true.

There were only nine aviation students in my class. The rest were aviation cadets. Columbus was a multi-engine school. It used AT-8, AT-9, AT-10, and AT-11 planes to train the students. I thought that the AT-9 was a great plane. It had small wings; therefore, it landed fast or hot. A lot of the students couldn't handle it, but I loved it. However, they later stopped using it because it was too much plane for the type of cadets they were getting.

In September, the "wheels" (Generals) decided to try an experiment. Word came down for the Director of Operations to select ten students, check them out in the T-6 single engine trainer, and send them to gunnery school to see how they performed. If it proved successful, they were to be put in A-20 attack fighters after graduation. My buddy Rosco Smith {"Smitty"} and I were two of the ten selected for the project.

The base brought in some AT-6 planes for our training. We had heard that the cadets in advance single-engine training were having trouble with the AT-6 because of its tendency to ground-loop, or swerve into a tight, uncontrolled circle after touchdown on a landing. We found out that in the single-engine schools, the instructors told the students that every landing was a potential ground loop, and if they ground-looped, they would be washed out of the course. It scared the students so much, it took longer to solo them than it did the twin-engine students.

By contrast, my instructor took me out to an AT-6 and said, "This is an AT-6. It flies just like the twin-engine trainers, as long as *you* fly *it*. If you let *it* fly *you*, you are in deep trouble." He showed me the major differences between it and the trainers we were using, went over the checklist with me, and said, "Let's do it."

The single-engine trainers had two cockpits, one behind the other. The plane could be flown from either cockpit. The student sits in the front seat; the instructor sits in the back. My instructor said that he would make the first takeoff and I was to follow him through on the controls. We took off and climbed to about 2,000 feet and went over near one of the auxiliary fields. He had me do different maneuvers and make two or three landings, then said, "Take me back to the base." After the landing, he said, "Take it into the ramp." We were scheduled for a four-hour period, and we had been flying for about an hour. When I parked, he got out and said, "Finish the period. Practice more maneuvers and landings so you can get the feel of it." I had heard that, at Craig Field (where they were training in the AT-6 planes), the students flew ten to twelve hours with their instructors before they were allowed to solo.

The next time the instructor and I went up, we went to the auxiliary field and practiced more maneuvers and practiced ground gunnery patterns without firing the guns. After about an hour, we went back to the base, and he said to finish the period and continue what we had been doing. Another hour elapsed, and I was getting a little bored, so I decided to do a little sightseeing. I saw a freight train, and I thought I would have some fun. I approached the train from the rear end, dived down as close to the train and as close to the ground as I dared. When I passed the engine, I could see the engineer through the window. I pulled the stick back and put the plane in a steep climb. When I did, the sudden change in altitude, coupled with the speed I was traveling, caused the engine to stop. I think the centrifugal force had stopped the gas flow. These planes were not designed to do that kind of maneuver. I was losing speed fast. I frantically pumped the throttle, and the engine started again—just before I was going to have to crash-land. It scared me so much, I never buzzed another train. I flew over near the auxiliary field and was doing some acrobatics, which we were not supposed to do, and Smitty called me on the radio and said he was going in to get some gas. I told him I would follow him in, as I needed gas also. As I approached the field, well above the traffic pattern, I saw one of the twin-engine trainers coming straight at me. I did a split $S$ (rolling over on your back and pulling through, like the bottom of an $S$) into the traffic pattern, went in, and landed. I thought it was a student who wasn't watching where he was going or didn't see me. As I was going to the line for gas, I got a call from the tower to report to operations. I was told the Director of Operations wanted to see me. It seemed that he was the pilot of the plane that was coming at me. He had seen me doing acrobatics and was trying to get my plane's number. He gave me ten one-hour tours, marching on the ramp with my parachute on my back. I had to complete them before I could graduate. It seemed rather stupid to me because doing maneuvers like that improved my skills as a pilot.

Word came down that we were going to Eglin Field, Florida, to fire aerial and ground gunnery. The ten of us, and our instructors, were taken to Eglin Field in a twin-engine passenger plane called a Loadstar. It was the largest plane I had been in.

The course in gunnery consisted of ground gunnery (firing at a fixed target on the ground) and aerial gunnery (firing at a large sheet of plastic being towed behind another plane). The ground gunnery pattern was a racetrack-shaped pattern at 1,000 feet and at a speed of 160 mph. You rolled out of the last turn approaching the target, slowed down to 120 mph, and dived, firing at the target as you approached it. After passing the target, you added power, climbing back to pattern altitude and speed for another pass. There were several planes in the pattern at the same time, equally spaced so that only one plane would be on the final run at a time. One of the students got the idea that if he slowed down, he would get more hits and get a better score. On one pass, he slowed down to 100 mph and just riddled the target. However, the plane behind him, flying at the normal speed of 160 mph, almost shot him.

In the aerial gunnery, each plane was firing bullets painted a different color. When the bullets hit the woven plastic and went through, the color would rub off around the hole, and they could tell which plane had fired the bullets. That way, they could tell each student's score. For safety, only one plane at a time would be in the aerial pattern. The same student who screwed up on the ground gunnery would wait until he was really close to the target, fire a short burst, pull up over the target, and go around for another pass. The tow ship pilot kept telling him on the radio, "Fire longer burst; fire longer burst." The tow ship finally told him to leave the target because they had other planes waiting. To show the tow ship that you were leaving the target, you would fly alongside and do a split S. All the tow target gunnery was done out over the Gulf, off the coast of Florida, so the bullets and the shells would fall in the water, not on the land, where they might hurt someone. As our "problem student" pulled out of the split S, he saw a school of porpoise below him in the water. He dived at them, firing what ammunition he had left in his guns, killing several of the porpoise. When he returned to the base, he was grounded, and when we returned to Columbus, he was released from active duty and sent home.

At the same time that we were in Eglin, the single-engine class from Craig Field was there for gunnery practice. Craig Field was training students in the AT-6 to be fighter pilots. They had done all of their training in the AT-6 and had about thirty hours in the plane, compared to our twelve hours. After the gunnery was over, we were told that the students from Columbus had out-flown and outscored the Craig Field Cadets.

The big day finally came. On October 9, 1942, I graduated from flying school as a staff sergeant pilot. My mother and father came from Sonora to see me graduate. I had the honor of being one of the pilots chosen to fly in the "fly by" formation that flew over the spectators before the graduation ceremony. After the fly by, we landed, changed into our dress uniforms, and attended the graduation ceremony. It was tradition for your girlfriend to pin your wings on your blouse after the ceremony. Since I didn't have a girlfriend, my mother did the honor. I felt 10 feet tall. I could see my dream coming true—and a whole new life for myself.

# Chapter V

## THE GOOD LIFE

During one of my physicals, the flight surgeon had decided that I needed an operation on my right foot to correct a short tendon to my big toe. Of course, it didn't matter that it had been that way since I had broken my leg when I was twelve. I had run the mile in track and played football and basketball with it that way when I was in high school. The flight surgeon said he would wait till after graduation to perform the operation. After the minor surgery, I was placed back on flying status, and I had a ball.

My classmates had shipped out to their new assignments. There was no new assignment for me. Since I was a "new breed" (a staff sergeant pilot), they didn't know what to do with me. The adjutant told me I could keep living in the cadet barracks, eating at the cadet mess hall, and should check with him about once a week to see if I had any orders to go to a new duty station.

The Director of Operations told the Operations Dispatch Section to give me an AT-6 anytime I wanted it. The only restriction was that I should stay within 1,000 miles of Columbus. At that time, my oldest brother was attending the Airplane Mechanics School at Biloxi, Mississippi. I would fly there, go to his commander, and get him a three-day pass. My brother and I would then fly around the country, sightseeing for two or three days. This was "The Good Life."

While I was at Columbus, I would go out with two of the sergeants who worked in the headquarters. The lower class, then all West Point cadets, was now the upper class and about to graduate. There were a few good guys among them, but they were the exception rather than the rule. Most of the West Point cadets were egotistical snobs. At about eight o'clock one evening, I was waiting at headquarters for my two friends to get through work. We were going out to eat. They had been working since early morning, trying to get the paperwork completed for the class graduation the next day. A West Point cadet came in and requested that one of the sergeants type a letter for him. The sergeant told him in a nice way that he was very busy trying to get the paperwork ready for their graduation the next day and didn't have time to type his letter. The cadet said, "I'll have you know I am General Yount's son, and I want you to type this."

The sergeant stood up, impressive at 6'-4" and about 220 pounds, and said, "I don't care if you even knew who your father was, you hit that door and don't come back." The cadet left in a huff.

My life of ease and cross-country flying continued. On December 2, I came back to Columbus to check in. I had a note on my door telling me to see the adjutant. He told me that orders had come promoting me to the rank of flight officer. Apparently, it was a "bastard rank" that some idiot had thought up to change us from enlisted status but not really to make us officers. Bars reflecting the rank of flight officer had not been made and were not available. I was told what they looked like and had to make my own. Also, there was no standard policy concerning the promotion of staff sergeant pilots. It depended on the base commander where you were stationed. Some bases promoted the staff sergeant pilots directly to second lieutenants. Others, like me, had to progress through the rank of flight officer. The adjutant told me that the only change in my situation was that I would have to move out of the cadet barracks and stop eating at the cadet mess hall. He said I could either move into the Bachelor Officers Quarters [BOQ] or get a place in town. I moved my things into the BOQ, turned in my laundry, got clean clothes, checked out a plane, and left.

While at Columbus, I witnessed an incident that pointed up some of the problems that the sergeant pilots encountered and was a preview of things to come that would affect our career from that time on.

One day a B-26 landed at Columbus. When a plane that large landed at a flying school, it was quite an event. A staff car was sent for the crew. A staff sergeant got out of the plane and was checking the tires and wheel wells. A major got out of the plane, followed by a captain. The aerodrome officer in the staff car asked the major if he was the pilot. He said, "No." He then asked the captain if he was the pilot.

The captain said, "No, I'm not. He's the pilot," pointing to the sergeant checking the wheel wells. The aerodrome officer told the major and captain to get into the car, and he drove off, leaving the staff sergeant pilot to walk to operations.

In another case, a base commander refused to let a staff sergeant pilot leave the base in the plane he had arrived in. The commander said, "No enlisted man is going to fly a plane on this base." The sergeant pilot had to call his commander and have him call the commander of the base where he was and tell him that the sergeant was a qualified pilot and to let him leave in his plane. It pointed up how ridiculous the flying sergeant program was. It was a stigma that followed us through most of our careers, affecting assignments and especially promotions.

The enlisted pilots got into the pilot program because they were in the Air Corps and loved to fly. Most of the cadets got into the pilot program to avoid being drafted. It was known at the flying schools where both enlisted pilots and cadets had trained that, on the whole, the enlisted pilots were far better pilots than the cadets. The commander of the Brady Texas Flying School stated that the first class of enlisted pilots at his field could fly circles around the cadet students whom he had known. I think that, later in their career, they continued to be better pilots.

# Chapter VI

## LADY LUCK

One day, when I returned to Columbus, I had another note from the adjutant. Orders had finally come, assigning me to Homestead Field, Florida. When I arrived at Homestead, I was assigned as an instrument flying instructor, a very easy assignment. I instructed new pilots on how to fly on instruments in bad weather. One week, I instructed four hours each morning, Monday through Friday. The next week, it was four hours each afternoon, and the following week, it was two hours each night. There was no weekend flying, and I had no other duties. When getting an assignment like that, one should treasure every minute of it because the bad ones will outnumber the good ones.

One afternoon, while we were practicing instrument landings at Homestead, I saw a crop duster plane dusting crops inside of our traffic pattern. I always looked for him as we circled the field so we wouldn't endanger each other. After several approaches, I couldn't locate the crop duster plane. Then I saw a large white area near the end of the field. In the middle of it was the wreckage of the plane. I reported the crash to the Homestead Tower. I found out later that the pilot of the crop duster plane had hit a power line at the end of the field and crashed. He had gotten out of the wrecked plane, walked out of the field and up onto the highway, and dropped dead.

Another incident at Homestead involved another instructor who, while instructing a student pilot in instrument landings, had had a stomach ulcer rupture. He was bleeding from the mouth, passed out, and fell forward on the control wheel. The instruments, like most of the cockpit, were all covered with blood. The student pilot held him off the controls with one hand and landed the plane with the other, without the aid of instruments. The student pilot was told he had done a good job. If that were to happen today, he would be given the Distinguished Flying Cross.

In February of 1943, I was flying low level over the Gulf near Card Sound. The surface of the water was flat, with no swells, and looked like a mirror. I lost the horizon, and we impacted the water with a tremendous noise and jolt. The impact tore off both engines. We were lucky that the plane floated, and neither the student pilot nor I was seriously injured. I had a large bump on my forehead, and the edge of one tooth was broken, but I was able to contact the Homestead Control Tower, and they had the Coast Guard pick us up. We were very lucky. I was to have many such encounters with Lady Luck in the years to come.

# Chapter VII

## OVERSEAS AT LAST

In March of 1943, orders came transferring me overseas. I was given sealed orders and was not to open them until I departed the States. My departure from the States followed a "cloak and dagger" scenario, almost like something you would see in the movies.

I was first taken to a small hotel near the 36 Street Airport in Miami. For the next three days, we had roll call twice a day and bed check at night. Then one night at eleven o'clock, I was told to get my gear together and be ready to leave. A covered truck came to pick us up. There were six of us, a Mr. Quintenilla (Ambassador from Mexico to Russia), two Intelligence types, and two other pilots. The truck took us to 36 Street Airport and stopped in front of a building, where our escort took us inside. We went through the building and out the back. There we got in a van. After driving all around Miami for about an hour, we stopped at the Pan Am Clipper Base. We obviously had taken a diversionary route from the airport. We were told to board the clipper that was parked nearby.

After takeoff and when we were past the point of "no return," I opened my orders. I learned for the first time that I was headed for a place called Chabua, India. The six of us were the only passengers on the huge clipper. The steward told us our first stop would be Belem, Brazil, in South America. They had some cargo for the base there. He said our next stop after Belem would be Natal, Brazil, where we were to spend the night. Later, he told us we wouldn't be landing at Belem because the cargo was a large casting and the radio operator had received word that there was no equipment at Belem that could offload it.

That afternoon, we landed at the seaport city of Natal—quite an experience for me, considering that the only city I had been in outside of the U.S. was a border town in Mexico.

Pan Am sent us into the city in staff cars and had rooms for us at the Grand Hotel. There was a large air base near the city, and the city was "off limits" to the U.S. military personnel stationed or transiting there. However, the restriction didn't apply to us since we were flying on Pan Am, so we had the run of the city. Although I could speak a little Spanish, it was very helpful to have Mr. Quintenilla as an interpreter. The people of Natal spoke a mixture of Spanish and Portuguese, which were quite similar.

We were told that the water in Natal was not safe and that we shouldn't drink anything that didn't have alcohol in it, so I went out and got myself a bottle of Seagram's VO. I met a girl in the hotel who said she was from the town of Recefe, down the coast from Natal. She had come to Natal to try to get a job with Pan Am. She was standing at the end of the hall by a window. I walked down there, and she was humming a tune. I asked her to sing a song and was surprised that she sounded just like Carmen Miranda, a popular Latin singer in the States.

The next morning, a Pan Am staff car took us to a very beautiful beach. There, we met some girls who were sliding down the sand dunes on coconut palm fronds. They showed us how it was done. I found out the hard way that on a high dune you can get up a lot of speed before reaching the bottom. I rolled quite a ways after I hit the bottom. We had a lot of fun with the girls at the beach.

The car came for us about noon. We were scheduled to take off at 2:00 p.m., and we had to eat first. After lunch at the hotel, we were taken to the plane. When the captain checked the plane before boarding, he found that the gas caps had not been sealed by the refueling crew. He refused to take the plane until all the tanks had been drained, refilled, and sealed. He told us that for all he knew, they could be full of salt water. It seemed that there had been some attempts at sabotage at the military base. The staff car took us back to the beach. We were happy to see that the girls were still there. The car returned for us at about 5:00 p.m. When we got to the Clipper Base, the Clipper crew was ready to go. We departed without incident.

We were told our next stop would be Fish Lake Liberia. During the night, the Clipper gave a big lurch, waking those who were asleep. We asked the steward what had happened. He said we had just crossed the International Date Line. It was the captain's joking way of letting us know. He acted as if the dateline were a bump in the sky.

Fish Lake was a rather large lake. Pan Am had a base on the shore. We were taken to the dining hall and treated to a wonderful lunch. After lunch, a truck took the six of us to Benson Field, where we boarded a C-47 "Gooney Bird" plane. One of the passengers on the plane was Martha Rae, the movie actress. She had been in Africa about five months and was going to Accra on the Gold Coast to put on her last show before going back to the States. That evening, we landed at Roberts Field in the jungle. It was the first time I had seen the jungle, other than in the movies. I was told that the natives working on the field were paid 30¢ a day. I couldn't believe people would work for so little pay.

We were given bunks in a long building. Behind the building was a deep river. In the jungle across the river, we could see fires and hear drums. We were told that the natives were having a feast for the Americans that night. I wanted to be sure that I had heard correctly that the feast was *for* the Americans, not *of* the Americans. The natives were to take us across the river in dugout canoes. One of the base officers told me the river was very deep and full of barracuda. We went to the "feast" and had quite a time. It seemed that the natives had an ulterior motive. They plied their visitors with "jungle juice," a strong drink made by fermenting rice and herbs, and then tried to sell them their daughters. There was a good feature in the sale. If you died or were killed, she belonged to your brother. Also, the price wasn't bad: The women were only $50. I don't think any of our bunch bought one because they couldn't take them on the plane and it would just have been a one-night stand. From what I found out, some of the men stationed there had already bought themselves a "sleep-in" maid.

Coming back across the river, one of the fellows who had had a little too much jungle juice stood up in the canoe, causing it to capsize, but he was able to swim to shore without being attacked by the barracudas. When he came into the barracks, he woke me up when he dropped his wet clothes, web belt, and pistol on the floor and crawled into bed. The plunge into the river and swim to shore had caused him to sober up.

The next morning, we took off and flew down the coast to Accra. As we flew along the coast, we saw two lines of natives, about 200 feet apart, starting at the water's edge and extending back up the beach. We couldn't tell what they were doing from the air. Later, we found out they were pulling in fishnets. We saw the same thing on the beach near Accra while we were there. The natives would take their nets out about 100 yards with their dugout canoes and drop them. With ropes tied at each end of the net, they would line up on the beach to pull in the net and chant, pulling in rhythm with the chant. When the net came in, it contained some of the weirdest fish one could imagine. Fish was apparently the staple diet of the natives.

We left Accra and flew to Madugri, Nigeria. The thing that I remember about Madugri was that a sergeant stationed there had bought a beautiful black Arabian horse for $50. He had raced it against many horses there and had won all the races. There was only one drawback: he couldn't take it back to the States. After Madugri, we went on to Khartoum, Egypt, arriving there in the evening. Operations told us that the city was "off limits" to transient personnel. However, one of the Intelligence officers called the base commander and got a staff car to take us into the city. We went to a club and had a delightful dinner. Afterwards, I saw my first whorehouse, run by the U.S. government. It had been created in an attempt to hold down venereal disease among the troops. It was run very business-like: First, you "signed in" at the office and were given a numbered slip of paper. All the girls were numbered, and you had to put the number of the girl you were with on the paper. The bar and girls were in a compound across the street from the office. There was a guard at the gate. You showed the guard the slip, and he let you in. When you left, you had to go back across the street, turn in the slip, and sign out. The girl's number was placed next to your name in the sign-out book. In case you got venereal disease, you were protected against courts-martial. At that time it was a courts-martial offense to get venereal disease. However, there was little chance of getting it from one of the girls—as a military doctor checked the girls regularly.

We left the next morning for Aden, Arabia. Aden is actually two cities— Aden and "Aden by the Sea"—divided by a range of hills and connected by a road that runs through a pass. While there, we were taken up a canyon and shown a series of dams where an ancient queen had workers construct a system for trapping rain water and bringing it through a conduit to the city for drinking water.

From Aden, we flew to a place called Salala, where we stopped long enough to get gas. Salala was out in the desert and was very hot. All the gas was in 5-gallon cans. It takes a lot of cans (160 to be exact) to fill up a C-47. There was a veritable mountain of gas cans there. After leaving Salala, we flew to Masira Island, off the coast of Arabia. We were told to stay out over the ocean on the way to the island and not fly over any land. It seemed that the Yemen people of Arabia had a pure race and that they wanted to keep it that way. After observing the American servicemen in "action," the Arabs were very smart to keep all Americans out over the ocean. Everywhere the Americans have been stationed, they have left a hoard of bastard children, many of whom—a large percentage, actually—were fathered by married men. These men would tell the girls they were going to marry them and would live with them, sometimes fathering more than one child. Then the men would rotate back to the States. In most cases, they failed to tell the girls they were leaving. Things got so bad in Vietnam that the commanding general put out a letter that the "tearful farewells" at the airport had to cease. We were told at Salala that a British crew had made a forced landing in that part of Arabia, and when the crew was found, they had been castrated and had their testicles sewn up in their mouths. I don't know if the story was true or just told to emphasize the point. Either way, it sure convinced us to stay out over the ocean.

After refueling at Masira, we departed for Karachi, Pakistan, which is on the coast near the Sind Desert. We were to spend the night there. I saw my first camel caravan, which was bringing goat and camel milk to the market. The next morning we departed for Agra, India, where the Taj Mahal is located. One of the Seven Wonders of the World, it is made of white marble, and we were able to see

*The tomb of Mumtaz Mahal, in the Taj Mahal. It is constructed of white marble, and inlaid with a floral design of semi precious stones.*

*The Taj Mahal, Agra India 1943 One of the seven wonders of the world. It was built in the 1600s and is engineering perfect. It was built by the Shah Jahan, for his favorite wife, Mumtaz Mahal. The dome is obscured by bamboo poles, used by the workers while cleaning and repairing the dome.*

it from a long way out. (I would be able to visit the Taj on a later trip to Agra.) That evening at the Officers Club , we were told horror stories about the kind of flying we would be doing when we got to our destination. We were told we would be flying from India to China over the "Hump," the same name that had been given to the Himalayan Mountains we would be flying over. I thought at the time that we were having our legs pulled. I was to find out later that, for the most part, the horror stories were true, and in some cases, the flying would be worse.

We departed Agra for Chabua, in the District of Assam, India, in the Bramaputra River Valley, about 75 miles from where the river came out of the mountains. At that point, the river is only a few hundred feet above sea level. The river starts in the mountains near Mount Everest (the highest mountain in the world) in the Himalayan Mountain range and flows from its source about 200 miles east through Tibet. It turns south about 50 miles to where it comes out of the mountains into the valley. From there, it flows west in a wide valley for about 200 miles, then south to the Bay of Bengal.

We finally arrived at Chabua, my base of assignment. It was not a large airfield and had one steel mat runway. The planes were parked in the dirt along the side of the runway. During the monsoons (the rainy season which lasts six months of the year), this area was mud. All of the buildings were made of bamboo, with thatched roofs. This was one of three airfields where Hump flights originated. More were being built. I would help open two of them down the valley. The operations building was near the runway, and the housing area was back in the jungle. We were taken to the housing area, where I was assigned a "basha"—as the bamboo huts were called. They were where the pilots lived.

*Part of a camel caravan bringing camel and goat milk to market.*

# Chapter VIII

# THE WELCOME MAT

I will never forget my first night in Chabua. There was a native village in the jungle nearby, and the natives were beating drums all night. As a boy, I had seen a movie called *Trader Horn,* which was about a safari in Africa, and the drums beating in the jungle spelled big trouble. I remembered that movie when I heard the drums. To make matters worse, the jackals started to howl, first on one side of the area and then on the other. Every time they howled, they would raise the pitch until it sounded like the high-pitched scream of a woman. Then the howl would break off into a hysterical laughing sound. What with the drums and the sound of the jackals, I was ready to leave there. The next morning, I learned what the commotion was all about. One of the natives in the village had died, which explained why they had beaten the drums all night. Somehow, the jackals knew. They also knew that they would soon have a feast.

The funeral procession came by the next morning, on the way to the burial ground, which was in a clearing in the jungle near the village. Leading the procession were four men carrying the body over their heads on a mat tied between two poles. These men were followed by the headman of the village, who was followed by a long single file of the villagers. The headman would chant something, and the villagers would repeat it. When they got to the burial site, they dug a shallow grave just deep enough to conceal the body, then covered it with dirt. As soon as they left, the jackals dug up the body and ate it. I was to learn later that when the natives in Burma buried a body, they put sharpened sticks in the ground all around the grave. They were sticking out in all directions to keep the animals from digging up the body. I went to the burial site near the base at Chabua several times later, and at times there would be parts of bodies scattered all around: an arm here and maybe a leg over there.

The Bramaputra River Valley was several miles wide and almost 200 miles long. It ran east to west. The east end of the valley where the river came out of the mountains was bounded by what was known as the "first ridge" of the Hump. To the south were the Naga Hills, which were inhabited by Naga headhunters. To the north were the Himalayan Mountains and Tibet. The river was very large, and during the monsoon season, it would sometimes completely flood the valley. Natives living there would build their huts on stilts so that when the river rose, their huts would stay dry. They had a dugout boat tied to the hut, and when the river rose, so did the boat. During floods, they would use the boat to travel to the jungle to get food. At times there would be very heavy rainfall during the monsoons. At a village in the Casio Hills about halfway down the valley, almost 400 inches of rain was recorded one year.

A narrow-gauge railroad ran along the river and brought supplies and barrels of gas for the bases and to be flown over the Hump. It was a common practice during heavy rain or fog to fly the "Iron Beam"—as the railroad track was called—since we had no radio navigational aids to guide us. At places in the valley, there were hills, and you would look out the window and follow the tracks so as not to hit the hills. We would also follow the river when it was not flooding, for the same reason. One day, a pilot was following the river during a heavy rain and didn't know there was a cable stretched across it. He hit the cable and crashed into the river. Although he had had a lot of mail on board for the men at the bases, as well as the bases in China, a lot of the mail was recovered. However, if the letters were written in ink, they were illegible. At that time, they had what was called "V-mail" ("Victory" mail). If the letters were V-mail, the water didn't bother them. Losing the mail really hurt the morale of the men whose letters were illegible because letters from home were a link to reality.

There were many kinds of wild animals in the valley. During the times when the river was flooding, they would move to higher ground along the sides of the valley. Wild water buffalo were a common sight along and in the river. Occasionally, we would see a herd of elephants or a rhino and would sometimes buzz them with the plane to make them run. The rhino was the only animal I saw that was not panicked by the low-flying plane. It would stand and jump up and down on its front feet as if to challenge the plane. We were later told not to buzz the elephants because they would sometimes run through the natives' village and destroy their huts.

I was wandering around the area of the base at Jorhat one day, when I saw a large grove of large bamboo that were 10 to 12 inches in diameter. As I approached it, I saw something in the top of the canopy. At first, I thought it was a group of monkeys. However, upon closer observation, I noticed that this was a group of large fruit bats hanging from the branches, upside-down. They would rest there during the day and forage for food in the evening and at night. I shot one so I could get a closer look at it. It had a wingspan of about three feet and was covered with a reddish-brown fur. Its head looked like a fox head, and it had sharp teeth.

One day I was walking in the jungle with a friend. I saw something sail from one tree to another. I told my friend to go around to the other side of the tree to make whatever it was move around to my side of the tree so I could get a shot at it, a tactic I had used when hunting squirrels as a boy. When it came around to my side of the tree, I saw that it was a very large flying squirrel. I shot it to get a closer look at it and to get the skin. It had a long tail that looked like a cougar's tail and had a web of skin between its front and hind legs. It also had a beautiful soft fur. I had heard of flying squirrels, but this was the first one I had ever seen. Its body was about two feet long, much larger than I had expected.

Several of the bases were built in or near tea plantations that were controlled by the British because, at that time, England controlled India. The women would carry large baskets into the fields and pick the young tea leaves from the bushes. When they filled the basket, they would put it on their head and carry it into the tea-wilting building. The building had several floors, and each floor had long

*Girl bringing a basket of tea leaves to the wilting shed.*

*A shed where tea is wilted before fermenting.*

multi-layer racks where the tea was spread out to wilt. After wilting, it was taken to a closed building (where the humidity was kept high) and fermented. The odor of that building would almost break one of the habit of drinking tea. After fermenting, the tea was put through dryers and then through rollers to break down the veins in the leaves. It was then ground and mixed with other types of tea to achieve the desired blend. After that, a professional tea taster would make several cups of the tea mixture to see that it was correct. If he approved of the mixture, the tea was put in 5-gallon square cans for shipping to tea companies.

In the east end of the valley on the north side of the river was a small airfield that was a fighter base with P-40 aircraft and about ten pilots. I had heard that the highest-ranking officer who had been on the field was a captain. The pilots were all second lieutenants, and the commanding officer was a captain. I had also heard that a general flew into the base one day and promoted all of the officers. They had a number of confirmed victories over the Japanese. The last time the Japs tried to raid the valley, twenty-nine were shot down. Two groups of P-51 fighters and the P-40s from Sadiya engaged the Japs, and the P-40 pilots got credit for 10 of the 29 kills. The easternmost transport base was Sookerating, which was in the east corner of the valley, near the first ridge of the Hump. The next field down the valley was Din Jan, which was operated by a Troop Carrier unit flying C-47's (Gooney Birds). The next base was Chabua. Next came Mohanbari, which was near the small town of Dibragar. Farther down the valley was the base of Jorhart. A little farther down the valley and across the river was the base of Tezpur. I started out at Chabua. Then when the base at Jorhat was finished, I was one of the pilots sent there to open that base in April of 1943. On April 27, I was sent to Gaya to transition into the Curtiss C-46. They needed more supplies in China than could be delivered by the C-47. The C-46 could carry more than double the load of a C-47. Ten airline pilots from the States were contracted to deliver ten C-46s to India, check out 10 crews, "sell" them the planes, and return to the States. I was one of the ten pilots chosen to be the first to transition in the C-46. I was given thirteen hours and ten minutes transition time in the C-46. However, only five hours were local training; the rest were trips to nearby bases. I "bought" the plane and flew it back to Jorhat. The next day, I was told to take the plane to Tezpur. I had been transferred there with some other pilots to open that base as a C-46 base. The C-46 was a nose-heavy plane, and if you didn't land with the tail low, you could nose over. While the pilots were getting used to the C-46, there were seven noseovers in one week. The base commander called a meeting of all the pilots. He told us that the noseovers had to stop and that the next pilot to nose over would be reduced to copilot. About that time, a dispatcher from operations came into the room and told the commander that Captain Keen, the chief pilot, had just nosed over. Of course, the commander had to change his decision. He couldn't very well reduce the chief pilot to copilot.

John Payne was the operations officer at Tezpur and had been an airline pilot prior to joining the Air Corps as a service pilot. Service pilots had an *S* on their wings. He could fly a Gooney Bird C-47 where most pilots couldn't drag a rope. I was told that he used to land on a polo field at a tea plantation and take the plantation owner's daughter to Calcutta with him.

One day, he took several pilots to Darjeeling up in the mountains. It was used as a rest area. Also, the British officers' wives went there while their husbands were on the front fighting the war. The pilots like to socialize with them to keep them from getting lonely.

There was no runway there, so John had been landing on a polo field, which had trees all around it. This time it was raining, and John overshot the field and had to go around. He slammed the throttles to the stop, climbed up, went around, and this time got it on the ground. He told the pilots to take the train back: he wasn't going there again.

I was waiting for him to bring the plane back, as I was scheduled to take it to Calcutta. When I got in the cockpit, I couldn't believe what I saw. The throttles were bent from his jamming them forward when he overshot the landing.

Another thing I remember about Tezpur was the large number of cobras in the housing area. They would come into the rooms at night. In the mornings you didn't just get out of bed: First you looked around the room and under the bed for cobras. In Calcutta, I had bought a young mongoose that was very adept at killing cobras, so she slept in the room with me, and I didn't have to worry about cobras in my room anymore. Sometimes the pilots would be sitting on the porch reading, and she would jump up into their lap and go to sleep. I named her "Little Audrey" and let her kill a cobra, which was interesting to watch because she would trot around it several times out of the reach of its strike, jump in, bite it on the back, and jump away. When she got the cobra weak, she would jump on it, biting it up the back towards its head. When she got to its head, she would crush and kill it. A mongoose has powerful jaws.

In mid-1943, the decision was made to expand the mission of hauling cargo to China. This was a result of Madame Chiang's taking a trip to the States, going before Congress, and complaining that we were not hauling enough supplies to China. Groups were being formed, and a contingent called the "Seventh Project" was sent over to man the new groups.

When the pilots of the Seventh Project came to Tezpur, the first time one of them saw Little Audrey, he thought he had found a wild animal, so he killed her.

# Chapter IX

## POTENTIAL FOR DISASTER

On July 21, I was sent to Sookerating to check their pilots out on the C-46. I stayed a month. One day I was sent to Mohanbari to pick up a load for China. The cargo was 8,000 pounds of 60 mm mortar shells. On the takeoff roll, at 60 mph, there was an explosion in the right engine, and the propeller went into over-speed or "ran away." I had two options: feather that prop and try to take off and put the fire out as I went around for a landing or try to stop before I ran out of runway and hit the jungle at the end of the runway. I didn't like either option because, to me, they both spelled disaster. I didn't think I could make it into the air and get around the field on one engine. Also, if I couldn't get the fire out, the gas in the right wing would explode and blow the wing off. On the other hand, if I couldn't get the plane stopped before running out of runway, when I hit the trees, the mortar shells would explode, and that would be the end of it.

At 60 mph, I had only seconds to make a decision. I decided that trying to stop would be the lesser of the two evils, so I throttled back both engines and applied full brakes. The end of the runway was coming up fast, and when I saw I was not going to get stopped before running out of runway, I applied full power to the left engine, released the left brake, and ground-looped to the right—off the runway. What helped save the plane was that the area between the sides of the runway and the jungle was deep mud (due to the monsoon rains). The plane spun around and stopped as it bogged down in the mud. By the time the plane stopped, the fire was out and my crew had jumped out and run. They had been afraid that the mortar shells were going to explode. It was funny because running wouldn't have saved them if the mortars had exploded because 8,000 pounds of mortar shells would have probably killed everything within a quarter of a mile.

# Chapter X

## SHORTAGES

I want to say that, at this time, I don't think the United States has ever been ready to go to war. They weren't ready for World War II, Korea, Vietnam, *or* Desert Storm. Our Defense Department doesn't appear to have the capability to plan ahead for contingencies. As late as Desert Storm, I saw trainloads of tanks and vehicles headed for the port en route to Desert Storm, but the problem was that they were painted in jungle camouflage green and were going to fight in the *desert*. As it was in India, we didn't have adequate supplies. The idiots in the Defense Department had decided not to send a lot of parts and supplies to India because they *"didn't want the Japs to know we were there."* It didn't matter that a Jap photo plane flew over the bases almost every day, taking pictures of us.

When a plane went down, a search flight was sent out to try to find the crew. Sometimes that took a lot of searching. If they went down in the jungle, which was usually the case, they would have to find a clearing so as to be seen. Due to the dense jungle, it would sometimes take several days for the crews to get together. Because we were not given any shoes to wear while flying, we had to wear whatever we could get. I was wearing low-quarter canvas shoes that I had bought at the market in Calcutta. Shoes like that would usually come off when your parachute opened during bailout. The procedure was to get together in a clearing, cut up your chute, and spell out on the ground the sizes of shoes needed. The search crew would write down the sizes and then had to return to its base. The pilot would tell the operations officer the number and sizes of shoes needed. The operations officer would write a letter to the Quartermaster Corps, stating that aircraft number 12345 had gone down, the crew had been found, and the following number of pairs of shoes, in sizes indicated, were needed. The quartermaster would then issue the shoes. The pilot would have to fly back to the area, find the crew, and drop them the shoes, food, and directions to the nearest village, where they could get help in getting out.

They say there is some good in everything. The only good thing about going down was if you survived and were found, you got a pair of shoes to wear when you resumed flying. I am surprised they didn't make the crews who walked out give the shoes back.

Shoes were not the only things we needed and were not issued. We didn't have warm flying suits, boots, or jackets. At altitude crossing the Hump, the temperature would sometimes drop to as low as -45° F. They had taken the heaters out of the planes because they were not sealed, and the barrels of gas we hauled leaked. They thought that some of the planes we were losing had blown up from the leaking gas fumes reaching the heaters.

The planes leaked when we flew through heavy rain, and we would get wet. As we climbed out, the rain would change to mush ice, then snow. Then it would

*On oxygen flying the Hump. At first, the mask did not have straps, and we had to use safety wire to hold it on. Later the mask had straps.*

be too cold for moisture. I have flown across the Hump with a coat of ice over both legs.

One day, the base received five electrically-heated flying suits. I got one of them, and it was great. The only problem was that the planes didn't have a temperature control, so when we plugged the suit in, it continued to heat up and had to be unplugged until it cooled down. Then we plugged it in again. The constant plugging and unplugging had to continue as long as the suit was used.

Another problem was the oxygen masks, which we had to use on every trip. The mask didn't have straps to hold it on, so we used safety wire to do the job, and the wire cut into our necks. Sometimes, when the wire hurt too much, I would loosen the wire, fly with one hand, and hold the mask to my face with the other. Some pilots took the tube from the oxygen regulator off the mask, put a cigarette holder on the end of it, held it between their teeth, and inhaled the oxygen. The problem with that was that they were getting pure oxygen, which I heard was bad for the teeth.

Another item was the food rations. We would have powdered eggs for breakfast and what they called "meat and vegetable stew" for lunch and dinner. The stew was just a cut above slop, and when we entered the traffic pattern in the mornings, coming back from China, we could smell the eggs. Some of them were green. One day, in 1943 the Command Section put a notice on the bulletin board (it was from Sector Headquarters, Office of the Commanding General) that said, "I note, with increasing alarm, the drop in the amount of supplies being delivered to China. Let's get on the ball and get the boys home by Christmas."

Someone had written below that: "We note, with increasing alarm, the drop in food supplies in the Officers' Mess. Let's get on the ball and keep the boys alive till Christmas."

We had been eating the "slop" a long time. One day, word got out that the mess officer had bought some ducks and we were going to have a duck dinner. Everyone lined up outside the mess hall early, waiting for the doors to open. They opened, and just as everyone sat down, the raid alarm sounded. As was the custom at that base, everyone headed out into the tea patches. There was no raid; however, the Japs had won a moral victory. When we came back to the mess hall, the duck was cold and covered with flies—not a very appetizing sight.

# Chapter XI

## ONE TURNING, ONE BURNING

On August 26, 1943, I was told I had orders to transfer to the base at Mohanbari. It was about 10 miles from Chabua. I was assigned to a basha with Rosco Smith, who had been in my class at Columbus, Mississippi.

One day, when I was flying a test hop in a plane they had been working on, the right engine caught fire. I turned on the fire extinguisher for that engine, and the flames died down but then flared back up. I saw I was not going to get the fire out, so I turned on the "bail out" alarm and told the crew to bail out. They asked me what I was going to do, and I told them I was going to try to get back to the runway. They said if I wasn't going to jump, they weren't either.

I called the tower and told them the plane was on fire and to keep the area clear so I could come straight in. I got the plane on the runway and got it stopped. We got out just before a wing blew off from an explosion of the gas in the wing tanks. Then one of the tires exploded from the heat, and the rim went clear into the jungle. A lot of men had gathered around, watching the fire. It's a wonder that the rim didn't kill some of them.

The crash truck came, but all it had was water (another shortage). They put water on the hot aluminum and magnesium, which caused more explosions. I had the crash truck back up to the cockpit window on the side that was not

*All that remained of the plane I brought in on fire.*

37

burning. I got up on the crash truck, crawled through the window, and started removing the radio equipment. I handed it out through the window to a man on the crash truck. I salvaged most of the radios. I would have gotten them all, but the heat was getting unbearable and I was afraid the other wing might blow. When the fire was out, all that was left of the plane was the tail and a part of one wing with the engine and wheel intact. The rest burned or melted. The fire burned a large hole in the runway and closed the field for two days. Afterwards, I walked back up the runway to where I had come in over the jungle and onto the runway. All the way to the end of the runway, there were large splotches of melted aluminum and magnesium, where the engine was melting as I came in. It was a wonder that the wing had not blown off before I landed.

One afternoon, while I was lying on my bunk, the raid alarm sounded, which was three Rapid rounds of the antiaircraft guns that were behind my basha. The standard proceedure in Raid alerts was for all pilots to report to operations.

When I got to the flight line, I found that it was not a raid afterall. The railroad that brought supplies to the base, also brought barrels of gas to be flown to China. The "Wogs"- as the natives were called— would roll the barrels out to a large storage area, and stand them on end. There was literally a sea of barrels. Some "brilliant" individual had decided to put dry grass on top of the barrels to protect them from the hot sun. When the old style wood burning train came huffing and puffing along the track, it blew sparks out into the storage area, setting fire to the grass covering the barrels. The raid alarm was used to get men to the flight line to roll barrels away from the fire to save as many as possible, but the barrels began to explode. It had gotten so hot, we had to leave the area. There would be an explosion, and a column of raw gas and barrels would go skyward. There would be a fireball, and out of the fire would come barrels of gas, with two jets of flame shooting out of the holes where the caps had blown off. It was a spectacular sight.

Because of my experience in the C-46, I flew most of the test-hop's at Mohanbari. One day maintenance called, and said they wanted me to test hop a C-46 with wing problems. Wing bolts had been coming loose, and they wanted me to take it up and see if I could loosen the bolts. A young second lieutenant Maintenance Officer said he would go with me. After take-off, I took the plane up to 15000 feet and leveled off. The lieutenant was standing in back of the pedestal, between the co-pilot and me. I rolled the elevator control nose down, putting the plane into a dive. When I got to 10,000 feet, at a speed of about 200 MPH, I spun the elevator control nose up. The plane shuddered, did a flat stall for another three thousand feet, before starting to fly. I looked over at the lieutenant, and he was white as a sheet. I told him, "You said you wanted me to see if I could weaken the wing bolts." I went back to the Base and landed. When they checked my wings, I had sheared over 100 bolts. The lieutenant never would fly with me again.

# Chapter XII

## MURPHY'S LAW

Murphy's Law states, "If anything can go wrong, it will." One day, I was flying the "Valley Shuttle," and I had taken some cargo into Jorhat and picked up 3,000 pounds of B-24 wing sections to take back to Mohanbari. Just as I broke ground on the takeoff, the right propeller started to over-speed. I put the prop in fixed pitch and continued to climb out. I thought I could use the decrease switch to reduce the rpm and go on to Mohanbari. I didn't want to spend the night at Jorhat. I leveled off at 4,000 feet and hit the decrease switch, but nothing happened. Since I only had 3,000 pounds of cargo, I decided to feather that prop and go on to Mohanbari on one engine. I flipped the right feather switch on. Again, nothing happened. None of the prop controls were working. I decided to throttle back until the rpm got within limits. I knew if I continued with the engine over-speeding, it could cause internal engine failure, which could tear the engine off the wing. I pulled the throttle back till I had climb rpm and headed home to Mohanbari. I leaned over and looked out the window at the engine, and it was doing fine. I told my copilot to watch the instruments on that engine, as it could overheat.

Everything went fine till we got about 10 miles from Mohanbari, when the rpm on that engine ran high. I pulled the throttle all the way back and trimmed the plane for single-engine flight. I called the tower and requested a straight-in

*Left to Right – Me, Joe Kennington, Alvie Hicks in Mohanbarri, India, 1944.*

approach. After I landed, the tower asked if I could taxi, and I said, "Sure." The tower told me to park on the left side. I gave the right engine some throttle, but I kept turning to the right, which I could not understand. I advised the tower that I could not taxi and requested that they send a tow truck. I was filling out the forms about the incident when I heard the maintenance crew hooking the tow bar to the tail. I climbed down the ladder and walked back to tell the maintenance crew what had happened.

The crew chief walked around to the front of the plane and said, "Lieutenant, come here. You don't have any right propeller." I went to see, and sure enough, there was just the splined shaft sticking out of the engine. The prop was gone. It apparently had come off about 10 miles out when the rpm had suddenly increased.

I took a lot of ribbing about landing without a prop and not knowing it. I told them, "Hell! That just shows skill."

One night about midnight, I was abruptly awakened by a loud roar and my bunk bouncing on the floor. My first thought was that we were being bombed. Then the antiaircraft battery behind my basha fired the raid alert. My helmet and gas mask were hanging on a peg on the other side of the room, and I tried to go get them but couldn't stand up, so I crawled across the room on my hands and knees, pulled myself up, and got them. When I crawled outside, I realized that it wasn't an air raid after all; it was a big earthquake. The ground was heaving in big waves, like the swells coming into the beach at the ocean—and I couldn't stand. The sound moved off down the valley. The earthquake didn't do very much damage at Mohanbari, but at Tezpur, about 75 miles down the valley, it cracked the slabs the bashas were built on. The earthquake was also felt in Calcutta, over 400 miles away. Not long after I returned to the States, the news reported another huge earthquake in the same general area that made the one that I was in insignificant by comparison. The news story said that the pilots who had flown there wouldn't have even recognized the terrain. Hills had risen up in the valley, and where there had been hills, they were leveled. Thousands of natives were killed, and the herds of elephants, as well as many other animals, were decimated.

As I mentioned earlier, the Japanese had a photo plane that flew over the bases in the valley while taking pictures. The Jap pilot would call on our frequency on the radio and tell us we were losing the war and should go home. He would be higher than our antiaircraft guns could reach—and above our altitude range. Someone finally had a solution. They took all of the armor plating off a P-40 and left only two guns. This reduced the weight enough so that it could reach the Jap's altitude. They took it to a small strip down the valley and waited for Photo Joe's return. When he was reported in the area, the P-40 pilot took off and spiraled up until he was at the same altitude. The next thing the Jap knew, there was a P-40 sitting on his tail. The P-40 pilot held his fire, making the Jap sweat. The Jap tried to get away but couldn't. Finally, the P-40 pilot gave him one burst of fire and shot him down. There was no parachute. The next issue of the *CBI*

*Roundup* (newspaper for the military) had a cartoon of the Devil standing by a pot with a fire under it and Photo Joe in it, with the caption "Photo Joe meets Dishonorable Ancestors."

It was SOP (Standard Operating Procedures) when a raid alert was sounded, for pilots to report to the line. Each aircraft was assigned two pilots, who would take it off and fly up in the hills to prevent being strafed on the ground. One day, when the raid alert sounded, two pilots got in a Gooney Bird to take it to the hills. They got one engine started but not the other. By then, the bombs were hitting the field. They shut the engine off, jumped out, and started for a trench. They looked up and saw a bomb coming down—right at them—tumbling end over end. They flattened out on the ground, and the bomb hit about 20 feet from them. They kept waiting for it to blow, but it didn't. It was found out later that it had hit flat, and with the kind of fuse it had, it had to hit nose-first to explode. After the raid, they were sitting on the porch of operations. One of them took out his 45 pistol and was fooling around with it when it went off, and when it did, their nerves, which had already been strained to the limit, snapped. They had to be sent to Calcutta for complete rest before going back on the Hump.

Whenever I had free time, I usually went into the jungle. I liked to watch the animals, especially the monkeys. They were always curious about me. One time, I spent the night in the jungle and decided to sleep in a large clearing. When I awoke the next morning, I was surrounded by monkeys of all sizes. Some of the females had babies holding on to them by the hair on their belly. They couldn't

*In the mountains of North East India, 1944, natives return from gathering food. They had killed a monkey and had some plants to eat.*

figure out what I was. As I watched them, I decided to try to catch one of the little ones. I picked out a likely target, jumped up, and gave chase. It got to a small tree and climbed it before I could catch it. I started shaking the tree to try to shake it down. I could shake one hand lose, but then it would grab the limb with the other hand. It finally got into a larger tree and got away. I was amazed at how well it could climb at that age. I loved the jungle and still do. I think it was Kipling who said, "If you have ever heard the jungle calling, you will always heed that call."

There were two things that I didn't like about the jungle: the mosquitoes and the leeches. Sometimes, when I would stop in the jungle, the mosquitoes would be so thick on my legs, I could hardly tell the color of my pants. The leeches were worse. I think they had radar. I could stop and look at the ground, and there would be hundreds of leeches standing on their tail ends and waving back and forth. They would then drop down and come looping straight toward me. Sometimes when I was walking through the underbrush, they would brush off on me. They usually bit me around the waist or inside my shoe tops. Sometimes I wouldn't know I had one on me until I felt a wet spot on my shirt. It would be a large circle of blood caused by a leech getting full and dropping off. When a leech bites you, it secretes an anticoagulant solution into the wound to keep the blood flowing. When they get full and drop off, the blood continues to flow for some time from the break they have made in the skin. After the flow removes all the anticoagu-

*Tibest natives rest on their way from Tibet to Indian villages to trade furs, crude silver jewelry etc for hats, clothing, knives and other items not available in Tibet. They make this trip once a year when the snow and ice melts in the passes.*

lant, the blood clots and stops the flow. However, if you later take a bath or go swimming in the river, the leech bites start bleeding again.

If we pulled a leech off, it would sometimes leave part of its mouth in the wound made by the bite. That would get infected, causing a jungle sore. I had heard that touching them with a burning cigarette would cause them to drop off without leaving part of their mouth under the skin. I tried and it worked, so I started carrying a lit cigarette to burn them off, which started me smoking. I continued smoking for about three years. I finally quit, though, which was the smartest thing I have ever done.

# Chapter XIII

## NOT MY JOB

My job was flying. I was trained as a pilot to do just that. However, one day I was contacted by Sector Intelligence, who had heard of my jungle experience and wanted to know if I would accept a mission into Burma. We had been losing a lot of planes crossing the Hump, and Sector Intelligence suspected that the natives were turning some of the crews over to the Japanese. The Japanese forces were as far north as the route we were flying from India to China. They wanted me to contact as many of the native villages that I could along the route the planes were flying, make friends with them, and ask them to turn our downed crews over to our side, instead of the Japanese. They said if I accepted the mission, they would provide me with an interpreter, some bearers to carry my supplies, and some "high" barter items to give the natives to win them over to our side. Although I could speak a little of their language, I would need an interpreter because there were so many dialects. In some cases, the people on one side of a hill couldn't understand people on the other side of the hill. It sounded like a great experience. Part of the time I would be behind the Jap lines, but that would just make it more exciting. I told them I would take the mission.

They cleared it with my commander and set up a flight to take me to a place called Fort Hertz, near the village of Putao, Burma, located in the upper valley of the West Irawaddy River. I was briefed that my contact would be a lieutenant with the Sector Intelligence working out of Fort Hertz. They provided me a complete medical kit with surgical instruments, bandages and sulfa drugs. The "high" barter items consisted of opium, salt, needles and colored thread. They also provided me with a large assortment of costume jewelry with colored-glass settings. These were items they thought the natives would go for.

The Gooney Bird took me to a small dirt strip near Fort Hertz. I didn't see any fort there. I just saw two or three buildings that were probably built by the British. The plane was met by the lieutenant and a sergeant who worked with him. The lieutenant said he had a place for me to stay while I was at Fort Hertz and he would take me into Putao the next day to meet my interpreter and four bearers whom he had arranged to go with me. That evening, he briefed me on the mission and showed me on the map where I would find villages along the route and where the Japanese were. He suggested a route for me to take that would get me through their lines, with the least chance of being intercepted.

The next day, we drove in the lieutenant's jeep to Putao, where I met my interpreter and the bearers. I told them I wanted to leave at first light the next day. The lieutenant had a trailer on his jeep and offered to take us as far as the West Irrawaddy river. The next morning, we picked up the natives in Putao and went to the river. I had provided each of the natives with a carbine and plenty of ammo. I also had a carbine. I got a native with a boat to take us about 2 miles down the river, where we would find a trail in the jungle, running east and west. We would be going east, the direction our planes were flying. We were to follow this trail east, through the jungle, and over the hills to the East Irrawaddy River. Beyond that river, the mountains were too high to cross. This route would get us through the Jap lines and would take us to the most villages. The trail, used by the natives, was easy to follow. This way, we would not have to cut new trail. The sound of cutting new trail would alert the Japs about our presence.

We went through this area very slowly, and in the area of enemy probability, we would move a few yards, stop and listen, move a few yards more, and stop and listen. We got through the area without encountering any Japanese. We did hear voices at one time, but they were some distance away, and we couldn't tell if they were Japanese or natives. We finally came to the first village along our route. The trip had taken two days. My interpreter asked the headman in the village if he had seen anyone like me before. He said, "No." He then asked him if he had seen any Japanese, and he said, "Yes." He then asked if he had seen airplanes flying over, and he said he had. The interpreter told him if he found anyone who looked like me and had come from one of the planes, he should take them to Putao, and he would be given money. I gave him some of the opium, and some of the villagers some of the other items. We moved on, as I wanted to get to the next village before nightfall.

We arrived at the next village just before dark. It was a small village consisting of three "long" houses. Their houses are made out of split bamboo with thatched roofs. The houses are built up off the ground on stilts (pigs are kept under the house) and are divided down the middle. There are five cubicles along one side, where five families lived. Out from the opening that went into each cubicle (they had no doors) was a firebox of sand in the floor. The firebox was about 3 feet square and the place the family did all their cooking. The smoke drifted up and out through the roof. With five families cooking at one time, the aroma certainly was interesting. The natives all smelled like smoked hams. There was an empty cubicle in one of the houses, and I was told that I could sleep there. That night, the men sat around a fire outside and smoked opium that I had given them. They had an iron ladle. They shaved some bamboo and toasted it in the ladle, held over the fire. They would then mix the toasted bamboo with some opium, which was a gummy substance, and put it in their pipes on top of a hot coal. The opium would melt and run down over the hot coal, and they would suck the resulting smoke into their lungs. Some of them got a little high.

The next morning before leaving, I gave the headman some more of the opium and the head of each family a gift of some of the other items. My interpreter told the headman that if they found any Americans from a crashed plane, they should take them to Fort Hertz, where they would be rewarded.

The villages were all interconnected by trails, which made our travel easy. The trail to the head village was well-traveled. They had heard through the jungle grapevine (communications system [drums]) that we were coming and had planned a feast for me. The village was on the bank of the East Irrawaddy River in a large clearing. At one end of the clearing were two large iron pots with fires under them. Nearby was a large log split into half. It looked as if it had been used for ages as a chopping block.They had brought a bullock of the Asian cattle family to the clearing near the split log and proceeded to kill it by plunging a spear into its heart through its chest cavity.The headman of the village then presented the bloody spear to me.To have the animal killed for a feast in your honor and then be presented the spear that killed it was supposed to be a high honor. I graciously accepted the spear.

The native men started to cut up the animal. They took the insides down to the river and washed them. In one of the pots, they had rice and herbs cooking; in the other pot, they put the insides of the bullock. Then they brought several baby pigs, burned the hair off them, and put them into the pot with the insides. They proceeded to cut up the carcass, and the head of each family was given a portion of the meat, depending on his stature in the village. The hide of the bullock was at least a half-inch thick. The head of each family would cut a 12-inch square of the hide for his meat. He would punch a hole in the center of the piece of hide, take a reed, tie a knot in one end, and run it up through the hide, with the flesh side up. He would then punch a hole in the pieces of his meat, run the reed through it, and slide it down onto the flesh side of the skin. When he finished, he would tie a loop in the reed. That was the way he would carry his meat.

The head of each family took a small piece of the skin, fixed it with a reed, took what he considered the choicest piece of his meat, put it on the reed, and presented it to me as his personal gift. Soon I had a handful of reeds of meat. I handed them to my head bearer to hold.

When I was young and my father, two brothers, and I went hunting and camping, we would take a green stick, put a piece of meat on it, and roast it over the campfire. I decided to do the same with a piece of the meat: I took a green limb, sharpened it, put a piece of the meat on it, and started to put it over the fire. When I did, all hell broke loose. You would have thought that I had assaulted the chief. They started pointing over to the edge of the jungle. I looked where they were pointing, and there was a platform about 6 feet high on top of four large bamboo posts. On the platform were skulls, bones, and I don't know what else. I asked my interpreter what was going on. He said that before I could put the meat over the fire, I had to take it over to the platform and ask the gods to get rid of the evil spirits. I knew that I was confronted with a situation that could cause big problems. The natives were very primitive—somewhat like children—and could be upset very easily. The last thing I wanted to do was upset them. For a minute or two, I just stood there, trying to decide what my next move should be. I thought of a ploy that might work: I stuck the stick in the ground. Counting on the natives' strong belief in Gods, I told my interpreter to tell the headman that my God would not let me do that. He seemed satisfied with that, and they went about getting ready for the "feast." They took the baby pigs (which couldn't have been over one or two days old) and the guts from the bullock, put them on

*Ceremonial stakes near the head village overlooking the East Irawaddi River in Burma.*

the split log, and chopped them up. For plates, they had cut some banana leaves into foot squares. They then scooped some of the rice from the other pot, put it on a banana leaf plate, then added about a handful of the chopped pigs and guts. That was the feast. They didn't eat any of the meat. They took it to their hut to dry for eating later. I had to have a very strong stomach to eat that, but I didn't dare not eat it if I wanted not to offend my host.

After we got through with the feast, the headman showed me how to shoot his crossbow. He was curious about my carbine, and I would have let him shoot it, but I was afraid the sound of the shot might attract a Jap patrol, and I certainly didn't want that to happen.

That evening, one of the women brought her daughter to me. The daughter had stuck a splinter in the top of her right foot, which had been wrapped in large leaves from the jungle. I unwrapped the leaves, and it looked as if gangrene had set in. I got my medical kit, took a scalpel, and cut away as much of the dead flesh as I could. I put some sulfa drugs on the wound and had my interpreter tell the woman to take her daughter to Putao, where there was a doctor. I warned her that if she didn't, her daughter would die.

The natives use bamboo for many things, such as carrying jungle juice, a strong wine made from fermented rice and herbs. The medium-sized bamboo, about 4 inches in diameter, has partitions inside about every 18 to 24 inches. The natives cut a section off just below a partition, and about 3 inches above the next partition. They punch a hole in the top partition, tie a reed thong to the top and bottom to put around their shoulder, fill it with jungle juice, and carry it with them when they go into the jungle. The next morning, the woman came back and brought me a joint of bamboo full of jungle juice as a gift for helping her daughter.

The time had come for us to start back, if we were to get to Fort Hertz by the time the plane was schedule to return for me. The headman told my interpreter it would be better for us to go down the river for about a mile, where we would find a better route back. He said two of his men would take us in their boats. I handed out gifts, and my interpreter told them to bring any Americans out to Putao and not give them to the Japanese.

The headman gave me a crossbow as a gift from the village. We went down to the river and departed downstream. We rounded a bend in the river, where we saw some natives working on a gravel bar. I stopped the boats to see what they were doing. They were sluicing gold in a very primitive but effective way. Out of bamboo, they had built a framework that was about 18 inches wide and about 8 feet long. It was about 5 feet high at the upper end and sloped down to the gravel. They had taken pieces of bark from a banana tree, which has hundreds of small pits on the underside, and overlapped them up the chute, with the pit side up. At the top of the chute was a large, loose, woven basket. They would dip up a large bamboo joint of gravel and sand and pour it into the basket. They would then pour water over the sand and gravel and wash it down over the bark, where the little pits would catch any gold. It was quite an ingenious way to sluice for gold. They would periodically take the bark off and retrieve the gold. They had a piece of small bamboo, about 3 inches long, with a hole inside about the size of a little finger. They had a cap for it made from a larger piece of bamboo. That is where they put the gold.

*Native girls playing a game in a village in northern Burma.*

We continued down the river until we came to the trail the headman had told us about. We proceeded as we had before and got through the Jap lines without any problems. We eventually came to a jeep trail—it wasn't a road—and had started along it when I heard a jeep coming from down the valley. It sounded as if it was in an awful hurry. When it came around the bend, it was almost on two wheels. It was towing a trailer, and when the driver saw me, he braked to a stop. He said that the Japs had overrun the front lines of the British troops and that they were falling back. He told me to get in the trailer and go with him. He said my interpreter and bearers would be safe. The Japs would not harm a native if they were not with a foreigner. I put my gear in the trailer, and we took off. There was a rope tied across the trailer that I could hang onto to keep from being thrown off. We would become completely airborne at times, when the jeep would hit a mound in the trail or top a rise. It was the wildest ride I had ever made. After getting to Fort Hertz, I was told that a jeep that was coming out after us had run off an embankment. They didn't know what had happened to the driver and his passenger. If they had been injured, the Japs had probably captured them. When we got to Fort Hertz, all of the springs were broken on both the jeep and trailer.

Down the valley from Fort Hertz was a Japanese stronghold on a hill named Supra Bum. The British forces had been trying to capture the hill for some time, but the Japs were well-dug-in with good fortifications. The British forces would pull back, and bombers would come in and bomb. By the time the British forces moved back up, the Japs were dug in again and would repel them. That is the area where the Japs had broken through. A month or two later, John Payne was returning from China in a C-87, a B-24 bomber which had been converted to a tanker to transport gas to China. When it was converted, they left the two 50-caliber guns in the nose turret. John thought he would help with the war effort, so he flew down across the hill at Supra Bum and strafed what he thought was the Jap positions. When he returned to Jorhat, he was informed that sector headquarters had received a message from the British command, asking that they kindly have the American pilots refrain from strafing their positions. It seemed that they had finally taken the hill from the Japanese the week before.

I debriefed with the lieutenant and went to bed early to get some much-needed rest on a bed. The next morning, the lieutenant informed me that the plane would not come for me for two more days. I decided to spend the time looking around the area, so I walked into Putao. There I saw a large Buddhist temple. I indicated to the monk at the door that I wanted to go inside. He motioned that I could. Inside, there was a large statue of Buddha about 8 feet high. I was told later that the head was solid gold and that the body was gold-leafed. There was a high metal crown on its head, with what appeared to be a very large diamond set in the front, with the other precious stones all around in the crown. There were two metal bands crossing its chest, and these too were studded with precious stones like rubies and sapphires and worth a king's ransom.

I followed a trail from the village of Putao out into the jungle. The trail led to a large clearing in the jungle, where I found the village burial ground. It was different from the one near Chabua. Here, the natives buried their dead a little deeper, and they put sharpened bamboo spikes into the ground, sticking out in all directions, to keep the animals from digging up the bodies.

I wanted to look around in the jungle to see what kind of animals were in that area. I followed an animal trail quite a ways into the jungle. There I found some temple ruins, all grown up with jungle growth. They looked to be centuries old. There was one large rock room that had a large idol of Buddha in the center. Around the room, there were niches in the walls where small statues of Buddha

*An idol in a Buddhist temple in Putao, northern Burma, 1944. The head is solid gold. There is a huge diamond in the head piece above the forehead. The metal bands across his chest are studded with precious stones, rubies, opals, saffires, etc.*

had been placed. Some were sitting Buddha's; some were standing Buddha's. However, over time, they had all fallen over. They appeared to have been carved from white marble and had been gold-leafed at one time. I had a shoulder bag with me, and I took three of the small idols and put them in the bag to take home with me. I wanted to add them to a large collection of artifacts I had collected in my travels.

Several years later, a visitor to my home was looking at my collection. She remarked that the small idols looked to be made of alabaster. I asked how one could tell. She said to shine a light through them, and if the light came through green, they were alabaster. I got a flashlight, and when I put the light behind the idols, it came through green. They were indeed alabaster.

I saw numerous monkeys, colorful birds, and two barking deer. When deer are frightened, they bark just like a dog. I returned to Fort Hertz and prepared for the return trip to my base in India. My mission was complete.

The plane came for me the next day, and I was looking forward to getting back to my base. When I got back to Mohanbari, I went to Sector Headquarters Intelligence to debrief on my mission. When I got back to my base, I went to debrief with my commander. He said he had a trip scheduled to go to Calcutta. He said I could take it and stay three or four days and relax, which sounded great to me. The next day, I took a Gooney Bird and went to Calcutta.

I stayed at the Grand Hotel in Calcutta, which was a very dirty city. I wasn't prepared for what I was to see there. The rice crop had failed, and there wasn't enough rice to feed the people. This was when I first learned that a large segment of the Indian population thought cattle were sacred and wouldn't eat any part of

*Snake charmers in the street in Calcutta.*

the meat. Their religion wouldn't permit it. Their beliefs were so strong that, even though they were starving to death, they would not eat the meat of cattle. That they were starving when there were fat cattle lying in the streets didn't make any sense to me. Many people in the States, as well as other countries, think they are very religious, but I wonder how many of them would give up their life before violating their religious principles.

After each meal, the hotel kitchen help would take all the scraps and left-overs out back and scrape them out onto the sidewalk, where the starving people would fight over them. The next morning after I got there, I went downtown. I had to walk around dead people on the sidewalk. I read in the local paper that over 600 people had starved to death in six weeks' time. I also read that in front of the local cinema one night, jackals had killed and eaten a starving woman. In the mornings, men would come around in carts pulled by oxen and pick up the dead. They would take them to a burning ghat, where they would be burned. I guess I had lived too sheltered a life. I didn't know conditions like that existed in the world.

# Chapter XIV

## ANOTHER DREAM COME TRUE

When I returned to Mohanbari, I was told that the "Old Man" wanted to see me. He told me he was putting me in charge of setting up a hunting camp in the upper end of the valley. The area was where the famous hunter, Frank Buck, did most of his hunting. It would be like hunting in a zoo. I loved to hunt and had killed my first deer when I was eleven. I had always dreamed of going big game hunting in Africa. This would be the best assignment I could imagine. This is how I got the nickname "BUCK".

The colonel said the hunting camp would serve two purposes: It would provide R&R for the men and much-needed meat for our meals. We had been flying fresh vegetables in from China, but we had no source for fresh meat. He said I would still have to fly once in a while, but I would run the camp and spend most of my time there.

I was assigned a weapons carrier that would take the men to and from the camp and bring the meat back to the base. The next day, I took a plane and flew up the river, looking for a good location for the camp. The area I selected was across the river in the Sadiya province. There was a barge that ferried people and vehicles across the river to the village of Sadiya. There was also an airfield near there where some P-40 fighters were based. A narrow road ran from the village of Sadiya up the valley and along the river but only went part of the way to the area I had selected for the camp. The rest of the way would have to be by dugout canoes. The next day, I took the weapons carrier, which was loaded with supplies for the camp, and went to Sadiya.

I was told to check in with the British Provincial Governor Williams of the Sadiya Tract. He lived in a very nice bungalow outside the village of Sadiya. In his living room was a fireplace with huge elephant tusks on each side that almost touched above the fireplace. The Governor told me that they were from a rogue elephant that he had to kill because it was endangering the natives who went into the jungle for food. Under his porch was another pair of tusks that he had taken for the same reason. He said that when the natives would report a problem with a dangerous or damage-causing elephant, he would kill it. He asked that we not shoot elephants, saying that there was a fist-sized spot between the elephant's eye and ear that you had to hit in order to kill the elephant. ( That's a very small spot to hit on an elephant.) He said we didn't have guns large enough to shoot elephants. If the elephant was wounded, it would become a killer and kill the natives who went into the jungle for food.

I asked Williams how long he had been there. He answered, "Two rains" and added he would be there for "two more rains." His assignment was for four rains. The natives speak in terms of "rains" (monsoon seasons) instead of years.

I explained to him what we would like to do: Since the base needed fresh meat, I would set up a camp in the upper end of the valley along the river. I told him that I had flown over the area and had seen a likely place to put the camp. He agreed that the area I had chosen was a good one. The only problem was that the road did not go all the way to that area. Part of the trip would have to be by boat. He said there was a lot of game in that area. I told him we would like to hire about 4 bearers. He said he knew a good "headman," whom he would have get 3 more bearers. They lived near the river in a small village called Sunpura, which was up the valley from Sadiya. The Governor said it would be safe to leave the weapons carrier at the head bearer's house in Sunpura. We would travel by boat the rest of the way. We followed the Governor to Sunpura to the head bearer's house. The head bearer spoke enough English that we would not have a problem communicating with him. He agreed to work for us and said he would get three more men and two boats.

I thanked the governor and told him we would check with him each time we went out and came back. The bearers got our supplies loaded them into the boats, and we set off up the river. I had selected an area just below a fork in the river, where it flows around a large island covered with elephant grass. There was a wide strip of gravel between the river and jungle where the river ran during flood stage. About 100 yards below where the river forked was a patch of jungle growth about the size of a city block. That was the place I had selected for the camp. I thought the location would give us some protection from elephants, tigers, etc. Below the camp area was another strip of gravel and another island of elephant grass. It was not a complete island; it was more of a peninsula bounded by the river on one side and a strip of shallow backwater on the other. The main jungle was down to the edge of the backwater.

*A Sambar deer I killed near the hunting camp.*

After we finished setting up camp, I took my rifle and went across the gravel to the main jungle. I wanted to try to kill a deer for camp meat, so I followed a wide elephant trail that wound around through the jungle. There were so many animals that it was like a zoo. I killed a barking deer and took it back to camp. That night, we could hear tigers calling in the jungle and elephants trumpeting as they crossed the river below camp.

The next morning, I got up early to go hunting. I was going back to the base that afternoon and wanted to take some meat with me. I took four bearers with me and went into the main jungle, the way I had gone the day before. I saw a bunch of wild hogs but couldn't get a shot. The bearers were talking and had scared them. I signaled to the head bearer that they had to keep quiet. A little later, I saw a sambar (large deer) about the size of a large elk and killed it. I gutted it and cut the head off. I told the bearers to cut a strong pole with which to carry the deer. I tied the legs together and put the pole through them lengthwise, and the four bearers carried it. I carried the head so I could have the antlers. After lunch, I took the bearers with me, and we went up the river to the grassy island where I had seen a herd of buffalo. I saw a nice fat yearling and killed it. I had the bearers get another pole from the jungle. I skinned the buff, cut off both the fore- and hindquarters and tied them to the pole. The bearers would not touch any part of the buffalo, as it was of the cattle family and their religion would not permit it. However, they would carry it on the pole. I carried the loins, as that is the best part of the meat. When I got back to camp, I wrapped the buffalo and deer meat in sheets I had brought from the base for that purpose. The bearers put the deer in the boat, and I put the buffalo meat in the boat, and we headed down the river. Two of the bearers stayed at the camp. They were to bring the other boat to Sunpura in two days. The four bearers would take four men and me from the base back to camp. I had decided that I would not have more than four hunters at the camp at any one time. We got the meat loaded on the weapons carrier, and I left for Mohanbari.

It was dark when I got to the base. When I took the meat to the mess hall, I cut a large steak off the deer and cooked it on the grill in the kitchen. It was the best meal I had had since I had come to India. The men were grateful to get the fresh meat for their meals as they were tired of eating the military rations of meat and vegetable stew.

I spent most of the next two months at the camp. I would go to the base about one day a week and fly the Hump or a valley run. The weapons carrier driver would shuttle back and forth from the base to Sunpura, taking men and returning with the previous hunters and the meat from the animals they had killed. Most of the meat was from yearling water buffalo, which were plentiful in the area.

One day, just after we had reached camp, one of the bearers came running into camp and said there was a tiger up the river from the camp. I told one of the hunters to get his gun and come with me. We ran up the riverbank and could see the tiger just below a patch of jungle growth, between the forks of the river. By the time we got across from that area, I could no longer see the tiger. I then saw him across the river from where he had been. He had swum the swift river. (I didn't know tigers could swim like that.) There was a strip of gravel about 100 yards wide between the river and the jungle, with an embankment about 8 feet high along the jungle.

The tiger was walking along, sniffing the gravel. It was broad-sided to us, about 150 yards away. I told my companion to get ready and, when I counted to three, to start shooting. In retrospect, I think it was a dumb procedure. I should have told him, "If I don't kill it the first shot, start shooting." When you count or wait for a signal, you get tense and don't shoot well. I missed. At the first shot, the tiger ran for the jungle and jumped the embankment as if it had been a low hurdle, and I couldn't tell if either of us had hit it. We went back to camp and took one of the boats across the river. I wanted to see if there was any blood where the tiger had crossed the gravel to the jungle, but we couldn't find any, so we must not have hit it. The tiger's tracks showed it was covering about 20 feet of ground each jump. It was a large Bengal tiger that stood about waist-high and had a huge head.

Another time, I left early one morning to hunt in the main jungle across from camp. It was very foggy, so visibility was not good. I circled through the jungle and came down a deep wash that came out onto the backwater below camp. Just as I got to the mouth of the wash, across the backwater from the peninsula of elephant grass, I saw two animals on the gravel below the elephant grass. At first, I thought they were deer. I couldn't see very well through the fog, and there was an embankment about 6 feet high along the edge of the grass. About that time, they jumped up the embankment into the grass. I said to myself, 'My God! Tigers!' I lay prone and waited. One of the tigers came back to the edge of the grass. I could see its big head and white chest. I aimed at its chest and fired. When I did, it let out a loud roar. I was so scared, my hair stood up. I couldn't have moved if a herd of elephants had come down the wash. At the shot, the tiger ran back into the tall grass, which in that area was about 5 feet tall. I decided to get the other hunters and bearers to help find the tiger. I had arranged a signal for anyone to use in case they needed help: Fire three rapid shots, wait a minute, and fire three more shots. I was about 400 yards from camp, so I fired the signal. The head bearer and two hunters responded to my signal. I told them about shooting at the tiger. My plan was to station one hunter on either side of the grass, one next to the river and the other near the backwater. If the tiger was not dead, it and the other one should come out of the grass at the lower end and one of the hunters should get a shot.

I sent the bearer back to camp to get the other hunters and bearers. I told him to bring enough pans and buckets for making noise. My idea was to line up across the upper end of the area and walk through, beating the pans and driving the tigers out the other end. One of the men said, "That's all right for you to go in there, Lieutenant, but who's going to take care of my wife and boy if I go in there?" I told him if he didn't want to go into the grass, he didn't have to. We lined up and went through the grass with no success. The tigers had apparently gone across the area, swum the river, and vanished into the main jungle on the other side. We never found any blood. I must have missed.

One day, there was a very unfortunate accident. One of the hunters was hunting buffalo with another hunter. They had separated, and he apparently wounded a large bull. From the evidence at the scene, the buffalo had charged him, and he had fired two shots at it with his pistol. The buffalo gored and trampled him. When the other hunter got to him, he was dead, as was the buffalo, which lay nearby.

# Chapter XV

## BACK TO THE WAR

One day, when I returned to the base, I was told I had to take a plane to Chitagong, which was in the southeast corner of India. It seemed the Japanese had moved out of Burma and were headed for a large British supply depot near the border in India. We were sending the transports there to drop supplies to the Allied troops that were trying to drive the Japs back into Burma. Sector Headquarters had also sent most of the P-51 fighters down there to provide cover for the transports.

Sector Intelligence had reported that there was a "war weary" Jap fighter squadron operating in the area. As usual, Intelligence was wrong. They weren't as weary as reported. They shot down the group commander and several of the other pilots. One day, our field was on an alert as the Jap's were raiding another field about 10 miles away. Near the runway were trenches where we could take cover. Suddenly we saw an aircraft approaching. As we got into the trench, one of the navigators stood near the trench and loudly proclaimed the aircraft was an American fighter plane. About that time, the plane opened fire, and a burst of bullets went right past the navigator, kicking up dust. He ran and did a broad jump into the nearest trench. He took a lot of kidding about the incident. We accused him of staying up all night and studying aircraft identification.

One day I was told to take some cargo to a small British field on the bay next to Calcutta. I was not in radio contact with them. As I came in to land, they started shooting red flares. I ignored them and landed. When I parked, a car came up fast, and a British officer got out. He approached me and said, "I say, don't go flapping your lips about wot you see 'ere. We don't want them to know, you know." It seemed they had just brought their Beaufighter planes into the area, and they didn't want the Jap's to know until they were put into action. I assured him that we were on the same side in the war, and I did not discuss military operations with anyone who was not supposed to know. When I came in to land, barrage balloons had gone up all along the waterfront. I had to wait until the British told them to take them down before I could take off. I stayed at Chitagong three days before going back to my base.

The Allies planned an invasion of Burma. Part of the plan was to take troops and supplies into Burma by gliders. It was a debacle. Some of the gliders landed in a swamp and crashed. Others ran into each other on landing because of insufficient space. More bad Intelligence. The Gurkah soldiers were used in the invasion of Burma and were commanded by British officers. I talked to a British officer who was with the Third Gurkah's. He said that the reason the Gurkah's were considered some of the best troops in the world was that they had no conception of being killed. He said the officers never led the Gurkah's into battle because they moved too fast. He said, "We just try to stay close enough to command them and keep them from going too far." I had heard tales of their daring exploits. In one case, a Gurkah took an injured British officer through the Japanese lines at night. He would put the officer down and scout the area forward. If he found a soldier sleeping, he would feel the brim of his helmet. The Japanese helmets were straight in front and the Allies' helmets had a brim. If the sleeping man's helmet had no brim, the Gurkah would cut his throat, go back and bring the officer forward, then repeat the procedure. He killed a number of Japanese on his way out to the Allied lines.

I read an Intelligence report of the North African campaign, in which the Gurkah soldiers were sapping (clearing) landmines at night. When it got light the next morning, the Gurkah soldiers had caught up with the Italians, who were laying the mines. They were removing the mines faster than the Italians could lay them.

An American engineering battalion was building a road into Burma from the upper northeast corner of the valley in India. This road started at a place named Ledo and was called the "Ledo Road." It was to join the famous Burma Road that came into Burma from China. One day, on the way to the hunting camp in the weapons carrier, I decided to take a trip along the part of the road that was finished. It was a hair-raising trip at times because it was so dusty, I could hardly see. I was told later that the truck drivers were given promotions, according to how many loads they hauled each day. They would drive as fast as they could, and if the truck in front of them slowed down for any reason, they would go around it. They were difficult to see, with all the dust they stirred up.

*The Ledo Road "Life Line" that took fuel for the epuipment, working at the end of the road. Burma, 1944.*

A pipeline had been laid along the road to take fuel to the equipment operating at the end of the road. The trucks that were carrying the pipe had long booms from the truck to the rear wheels on which they carried the pipe. The road had many hairpin turns, and once or twice, I met one of these trucks on a turn. On one occasion, I met a truck as it came around a turn. The rear wheels cut across, trapping me next to the embankment. The driver stopped, and I had to back slowly as he inched forward until the road straightened out and I could get out of the trap.

Where the road crossed the first mountains, they followed a pass. They had blasted off the ridges and filled in the gorges. At one place they blasted into a cave of huge pythons. The blast killed them, and the workers put them in a weapons carrier that was full of dead snakes.

I stopped and watched a sergeant on a bulldozer pushing boulders off the side of the mountain where they had blasted for the roadbed. He was working very fast and would push a bunch of boulders over the edge of the cliff, then back up fast before the dozer went over with the boulders. At the edge of where he was working was a barrel of fuel oil that he needed to move out of his way. Instead of getting off the dozer and moving it, he drove up to it, set the dozer blade on top of it, laid it down, pushed it out of his way, set it back up with the dozer blade, and continued pushing boulders. He was good at his job.

# Chapter XVI

## THE HUMP

The Himalayan Mountains start west of Mount Everest, the highest mountain in the world, and extend east through Tibet, into China, and south into Burma. That part of the mountain chain that lay between the bases in upper Assam, India and the bases in China to the east was dubbed the "Hump." In the early period of Hump flying, there was no designated route to fly. We were given a plane number and a destination in China, and we got there any way we could. The weather was such that there were very few days when the Hump was open or clear. It was acknowledged that the weather on the Hump was the most severe in the world, changing in a matter of an hour, from clear to very severe weather, with high winds and heavy ice, accompanied by violent up- and downdrafts. The minimum safe altitude on a direct route to China was 17,000 feet. To the north, it was much higher. The average terrain in that area was 20,000 feet of brown rock and glacier lakes. It was said that the clouds on the Hump had rocks in them. Many planes were lost by flying into clouds that had rocks in them.

I will never forget my first trip over the Hump on April 10, 1943 (I was 23 years old). A general said we were accomplishing the impossible, primarily because we were too young to know the impossible. The normal procedure was to make three trips as copilot for familiarization and then take over as first pilot.

My pilot on my first trip was Pat Patterson. I think it was his first trip as first

*The Hump.*

pilot. The trip over the Hump was beautiful. The Hump was open, with just a few clouds. We had been told that we would be given an Air Medal for each twenty-five trips across the Hump—and the Distinguished Flying Cross if we lasted fifty trips. Looking at the Hump that first trip, I couldn't imagine why they would give medals for that kind of flying. I would soon find out why.

We were on the ground about two hours before heading back to India from Yunnanyi, China, the most westerly base in China. There was a mountain peak just west of the field named Mount Tali, which was the beginning of the Hump. We circled a few times to gain altitude before heading past Tali. The normal wind on the Hump was from the Northeast and averaged 60 mph. Going over, we had the normal Northeast wind, so Pat corrected to the right to offset that wind drift. Just past Tali, we went into the clouds. At first, the ride was relatively smooth. Then we started picking up ice on the wings and windshield. The only navigational aids on the Hump at that time were fifty-watt radio homing stations on each end of the line. They were so weak, they couldn't be picked up over 20 miles from the station, so we were not aware that we had a wind shift to the southwest. It was blowing us north, and our wind correction was making it worse.

We had climbed up to 19,000 feet. There were mountains to the north of our route that were as high as 25,000 feet. Pat had calculated our flying time to Chabua as two hours and forty-five minutes, considering the Northeast wind we had going over. That time would put us well over the valley. We were now in heavy snow and ice. When we reached the estimated time for the flight, Pat started letting down. We still could not get the homing station at Chabua. I was pumping de-icer fluid on the windshield so we could see out. We had let down to about 9,000 feet, thinking we were over the valley. I happened to look out of my side window, and just past the wing was the side of a mountain. I said, "Pat, let's get the hell out of here. We're in the mountains. Turn left. There's a mountain just past the wing." He looked out his window, and I saw his head tilt back as he looked up, and I knew what he had seen.

Fortunately, we had let down in a canyon. There was a river below us, and Pat said, "That river is almost on course. I'll follow it out to the valley." As we continued to let down, I looked ahead, and the canyon was full of clouds.

I said, "Pat, you can't go into those clouds. If that river turns, we will hit the mountain."

He said, "I can't turn."

We went into the clouds, and when we came out of them, there was a mountain right in our face. Instead of going downstream, we had been going upstream, and the river forked. I turned around and yelled to the radio operator to brace himself for a crash. I couldn't see any other possibility.

Controls of a plane have two rudder pedals. If you push one and hold the wings level, you will skid in that direction. Pat said, "Tell me when we're up against the mountain on your side." He skidded to the right.

When we were up against the mountain on my side, I said, "That's it." He pulled it up into a stall and did a wing over to the left. When we headed back

*Rocks that kill pilots.*

down the canyon, he added climb power, and we went back into the clouds, snow, and ice. I couldn't listen for the homing station because the snow static makes a loud roar in the headsets, and we had taken our headsets off.

We finally came out on top of the clouds at 20,000 feet, and there were mountain peaks sticking up through the clouds on both sides of us. Fortunately, the river didn't turn. We knew we were way up north, but I didn't know exactly where. I put my headset back on and called Chabua for a bearing. I gave them a long count so they could get a fix on us. They gave us the heading to fly to get to Chabua and said to call back in five minutes to get a wind correction.

All this time, I had been tense but relatively calm. I have always had a knack for staying calm during great danger. However, when it was all over and we had the new bearing, I snapped. I started shaking so hard, I couldn't sit in the seat. I had to get up and walk to the back of the plane. When we got to Chabua and went into operations, I took off my wings, slid them across the counter to the operations officer, and said, "You can have these. I want a ground job." It was the only time I have ever wanted to stop flying. I had been so scared, I wanted to quit. He just laughed and said I was scheduled to fly the next day. They were laughing about the radio operator. Because we had taken our headsets off, we didn't know that the radio operator had been crying over the radio, "For God's sake, someone give us a bearing! We're going to crash!" They thought that was funny, but if they had been there, they wouldn't have thought it was as funny. Afterwards, Pat got so drunk, he shot and killed his bearer, who had taken care of his basha.

There was a captain there who couldn't safely fly instruments. Because I had been an instrument instructor in the States, when I completed my three familiarization flights, instead of letting me take over as first pilot, they scheduled me to fly with him as his copilot. They were afraid he would kill his crew and lose a plane.

I had made three trips with this captain. On the fourth trip, we were in heavy snow and ice at 21,000 feet, and I was flying. The right engine quit, and we lost altitude down to 15,000 feet before we could get it started. When we got in the clear, I turned it back over to the captain.

When we got to Kunming and started our approach, the right gear didn't go all the way down and then wouldn't come back up. The indicator showed the gear was not safe. I told the captain to fly past the tower and let the tower operators look at it. They confirmed that the right gear was not all the way down. We climbed back up and tried to pump up enough pressure to force it down, to no avail. We tried to air-swing it (dive with a fast pull up to cause centrifugal force to make the wheel go down) without success. By then, a Curtiss Wright tech representative had been told about the problem and had gone to the tower. He started telling us things to do to get the gear down. I kept saying, "We tried that. We tried that." I told him that the only thing left was to bounce the plane off the runway on that wheel and try to drive it in place.

He said, "That's only as a last resort."

I told him it was the only thing left to try. If it would come back up, we could make a belly landing, but with one down and one up, we would surely crash.

The captain asked me to get in the left seat and fly it. I flew down the runway at 150 mph, rolled the right wing down, driving the right wheel into the runway.

*Waiting for the "Jap's."*
*Flying Tiger P – 40 fighters on the field in Kunming China, 1943.*

When it bounced, I added climb power and went around. I flew by the tower, and they said it appeared to be all the way down. However, the gear indicator still gave an unsafe indication, and the warning horn blew.

I had 24 barrels of gas on board. There was a large lake near the field. I told the tower I wanted to throw my load into the lake. The tech rep asked how much hydraulic pressure I had. I told him, and he said with that much pressure, the gear would stay down, even if it wasn't locked.

I told him to keep other planes out of traffic, and I would land. When I was about to land, the tech rep said, "I've told you all I know. If you want to throw your load in the lake, go ahead." He knew that I was committed to land. Also, there was a good possibility that the wheel would fold and cause a crash. If that happened, the barrels of gas would make a good funeral pyre. As soon as he said it, I knew he said it to clear the base commander of any blame if we crashed with the gas on board. He could say, "We told him to throw his load into the lake." I was furious. I landed on the left wheel, let the right wheel down slowly, and rolled to a stop without using brakes. A jeep came out from the tower to where I was stopped on the runway. It was the tech rep, a small man who ran under the wing to look at the wheel strut. When he did, I decided to get even with him for what he had said. The runway was surfaced with small gravel. I advanced the right engine to full power, blasting him with small gravel from the runway. Later, one of the ground crew told me I had almost blown the tech rep away.

What had caused the problem was a small metal box that contained the switch for the gear warning system. It had come loose, dropping down and jamming the gear. When I bounced the wheel on the runway, the gear locked into place, but in doing so, it had destroyed the switch, along with the electrical circuits. That was why I had still gotten an unsafe warning. If I had crashed, they would have said it was "Pilot Error." The people at Kunming were quick to declare "Pilot Error" if there was a crash. When I got back to my base, I told the chief pilot that I would not fly with that captain again. He not only couldn't fly instruments, he wouldn't fly the plane, which was his responsibility when there was an emergency. The chief pilot said I could start flying as a first pilot.

On another occasion, I arrived at Kunming just at dark. The runway was northeast/southwest. On the southwest end was a big valley with the large lake. On the northeast end was a small village, beyond which were hills that stretched across the approach to the runway. You came in over the hills, then the village, and onto the runway, which started just beyond the village. When I called for landing instructions, the tower said to land to the southwest. They advised me that there was a large thunderstorm over the hills on the approach. I could see constant lightning in that area, so I asked the tower for permission to land to the northeast, downwind. They cleared me to land to the northeast. After I landed, a very heavy rain moved over the field. I parked and waited for the rain to stop. The rain moved away from the field, and we were standing by the plane, waiting for transportation. The plane following me was on the approach to the runway, from the northeast. When I landed to the northeast, the Chinese ground crew moved the portable floodlights to the southwest end of the runway. The lights are used to light the approach end of the runway. They turned them on as the plane approached the other end of the runway, blinding the pilots. A steamroller had been left in the center of the northeast end of the runway where some construction was in progress. The tower failed to mention that to the pilot. Just as the pilot touched down, he saw the roller, pulled up over it, stalled, and came down hard. He hit the runway, bounced to the right, and crashed into a building. He had 24 barrels of gas on board, and when he hit the building, they exploded. The entire crew was killed. The crash was "investigated" by the base flying safety officer. His report stated that "all" planes had been landing to the northeast and that the pilot had landed contrary to normal traffic, causing the accident. 100% Pilot Error. I was the only pilot to land to the northeast. Also, there was no mention of the roller on the runway. The report cleared the base commander of any responsibility for the accident. The flying safety officer was sucking up to the base commander.

I flew into Kunming one night. The plane I was flying was using excessive fuel. The base had put out a memo stating that if you had on board more gas than they thought you needed to return to your base, they had the right to drain part of the gas out before you could leave. They had computed the normal gas consumption from Kunming to each of the bases in India. With the head winds I would have going back to India, I didn't have enough gas to get to my base with any reserve. To compound the problem, at that time of year, the upper valley was

*Flying the passes on the Hump, in a heavily loaded "Gooney Bird." Minimum safe instrument altitude on a direct route was 17,000 feet.*

usually fogged in the early mornings. Often we would have to go a long way down the valley to find a base that was open. I told the operations clerk that I wanted 200 gallons of gas because my plane was using excessive gas. He showed me the memo. I told him I didn't care what the memo said. He called the operations officer, and I told him what the problem was. I remember that he had a speech impediment, in that he would put an *h* on the end of most words that

*Flying below the peaks on the Hump. There is some type of man made structure in the lower left corner.*

ended with an *s*. He said, "I've got 10,000 hoursh, and you can't tell me any C-46 burns that much gash."

I said, "Major, the only thing I'm telling you is, I'm not taking off till I get the gas, or it gets daylight."

He pointed to the dispatch board and said, "Two pilots just took off without extra gash."

I said, "Major, do you know those pilots?" He said he didn't. I said, "I do. They are new on the Hump and don't know better." As it turned out, one of them ran out of gas and had to bail out.

Finally, the major said, "Sergeant, give this man the gash he wants and get him out of here." When I got to the valley, I had to go clear to Jorhat before I found a field that wasn't fogged in.

One of the biggest problems we had on the Hump was the severe turbulence. The up- and downdrafts were horrendous. When you hit the downdraft side, you would have to add climb or max power and pull the nose up, and you would still be going down 4,000 or 5,000 feet a minute. You would then hit the updraft side, and you would pull the throttles all the way off, drop the landing gear, and put the nose down. You would still be going up 4,000 or 5,000 feet a minute. The ones on instruments in the clouds that got below minimum safe altitude in a down draft had to bail out.

One of the pilots got into heavy turbulence one day. It flipped him on his back and put him in a spin. He dropped 10,000 feet before he could get control and get the right side up. Fortunately, he was in a canyon. When he returned to his base, they inspected the plane and found that the main spar in the right wing had bent from the stress.

Most of the trips, we carried 24 barrels of gas, which would be tied with ropes. Sometimes the violence of the up- and downdrafts would break the ropes and throw the barrels all over the cabin.

Sometimes on the return trip, we would have 60-pound "pigs" of tin or tungsten. They would be stacked down the center of the cabin floor and tied down with ropes. In the turbulence, the forces exerted on the ropes would break them, throwing the bars all around the cabin. Occasionally, it would throw some of the bars through the side of the plane. Some planes would return to India with large holes in the sides of the plane, looking as if they had been hit with antiaircraft fire.

On the majority of the trips over the Hump, we encountered snow and ice. Sometimes the ice would get so thick on the props, it would break off and come through the sides of the fuselage. It would be very cold — at times as cold as -45°F. On more than one occasion, I had the oil in the oil cooler congeal and, unable to thaw it, had to shut down that engine and continue on one. Another problem was the loss of power on the earlier models of the C-46, which baffled me. We would be flying along, and without any warning, one or both engines would start losing power, and we would start losing altitude. What was baffling was that all the instruments would be indicating normal operation. I would try everything to try to correct it: put the mixture controls in the rich position, apply carburetor heat and de-icer alcohol (in case it was carburetor ice), and advance the throttles to try to get more power. Nothing worked. We would lose a few hundred or a few thousand feet. All of a sudden, the power would come back, and we would start to climb. As in downdrafts, if we were in the clouds and had reached minimum safe altitude, we had to decide to stay or bail out. Some crews had to bail out. It was so mystifying that a rumor got started that the Jap's had a ray gun that was robbing the engines of power. The problem was finally solved by Wright Patterson Air Depot. On test stands, they found that a fuel line was routed too close to the engine manifold. Under certain conditions, the fuel would start to boil, causing leaning of the fuel mixture and a loss of power. They re-routed the line, and it worked fine, but not before we lost many planes.

A number of planes were shot down by Jap fighters. There was a Japanese airfield just south of the direct route over the Hump. The Jap planes would come into it in the evening, refuel, and (the next morning) try to shoot down the un-armed transports. We didn't have escorts. I think the most they shot down while I was there was ten, in December 1943. When the Hump was open, we would fly up north to avoid interception by the Jap pilots. One day, while we were going back to India, the radio operator saw two Japanese fighters coming up towards us. They usually stayed low, and when they saw a transport, they would climb up under it and try to shoot it down. We were approaching the last ridge of mountains before getting to the valley. I was at 25,000 feet, so I put the nose down, added power, and headed for the valley. I called the Fighter Net radio, advised them that I was being intercepted, and asked for any of our fighters in the area to intercept me and provide me with an escort. I was advised that there were no fighters available. When I passed through 20,000 feet, I glanced at the airspeed instrument, and the needle was up against the stop, at 360 mph. The wind was screaming around the plane. I literally ran off and left the Jap fighters. After I got back to my base, I computed my speed, based on the temperature at 20,000 feet, and discovered I had been doing about 500 miles an hour. It was probably a record for a transport aircraft, and I am positive it was a record for the C-46. I kept climb power on until I reached treetop level above the valley floor.

Crossing the Hump at night, we didn't turn on our lights, for fear of Japanese interception. It wasn't unusual to be flying along and see the shadow of another plane as it passed me going the other way. As I said before, we were not assigned an altitude to fly: we flew where we wanted to fly. There were times when a plane going east and a plane going west would fail to return to their respective bases. I'm sure that in some cases, they hit head-on. They later designated routes and altitudes for the flights. They also designated a plane as a weather flight. During bad weather on the Hump, each hour a plane would be designated as the "weather ship." They did this because many inexperienced pilots were coming to India and were turning back when they hit bad weather. They would always put an experienced pilot on the weather ship. If he got through, he would keep the Hump open. If it was too bad, he would close the Hump for an hour until the next weather ship. This continued until they got a stupid, young colonel over there who wanted to impress his superiors. One day, he was going to China in a B-25 and, because he had guns on his plane, flew south. The weather was always better south, and the terrain was lower. He didn't worry about Jap interception because he had guns. He asked his radio operator if he had heard from the weather ship. He said he had. The colonel asked him what the weather ship did, and he said, "They closed the Hump." When the colonel got back to India, he said there would be no more weather ships and that every plane would be a weather ship, adding that he was going to move the cargo to China, if he had to build an "aluminum highway." The aluminum highway would be the fuselages of crashed airplanes. He almost succeeded. At the same time that the weather ship closed the Hump, another plane went through, and when it got to China, it was Class-26 (not reparable). Hail had ruined the plane.

We were given a list of code words to use to prevent the Jap's from knowing what we were saying over the radio. One day, I was going into the field at Yunnanyi, when the tower said, "Plane 683, Red Wing, Red Wing." I told the copilot to look up Red Wing, and I continued my approach. Again, the tower said, "Red Wing." The copilot was running his finger down the list of code words, saying, "Red Wing. Red Wing." I told the copilot to put the gear down.

He replied, "Gear down."

Upon seeing the gear go down, the tower broke radio code and said, "683, get the hell out of here. We're under a Red Alert."

*A glacier lake at the 20,000 foot level on the Hump in southern Tibet. When the Hump was clear, we flew north to evade Jap interception.*

I called, "Gear up!" and dropped down just above the rice paddies and headed north for the mountains and safety. When the tower called and gave me the "all clear," I returned and landed. The only damage to the runway was on the approach end. The Chinese were working on that end of the runway. They had a rock roller, about 4 feet in diameter and 6 feet long, with a spindle on each end. About fifty Chinese would get into a harness and pull the roller to pack the gravel on the runway. A bomb had hit in the middle of the workers, and there were bodies and parts of bodies stacked at the end of the runway.

One day, one of the pilots going into Yunnanyi lost an engine. Then the other engine quit. He picked up the mike and said, "I've just lost both engines. Clear the area. I'm going to make a "dead stick" landing and save the government $300,000 worth of airplane. The radio operator wrote in his log, "Going to save government three hundred thousand dollars." The pilot forgot that with both engines dead, he could not put the gear or flaps down, which would slow the plane down, slow enough for a landing. When he got to the runway, he was going about 150 mph. The radio operator looked at the airspeed, then out the window and saw the tower go by. He took his pencil and rubbed out the "three hundred thousand dollars" in his log. He knew that wasn't true. They made a belly landing in a rice field beyond the runway.

On another occasion, a Major Rice going into Yunnanyi forgot to put his gear down. When he hit the runway and skidded to a stop, both engines and the belly of the fuselage were ruined. They later towed the fuselage to a spot near the operations building and made a coffee shop in it. It was named "Rice's Coffee Shop."

On one trip to Kunming, the weather was worse than usual. When I got there, planes were stacked every 500 feet from the ground up to 15,000 feet. As a plane would land, everyone would let down 500 feet. I joined the stack and had worked my way down to 10,000 feet. Other planes had joined the stack above me. About the time I reached 10,000 feet, a plane above me in the stack had run out of gas and said he was bailing out. That meant that his crew and the plane had to come down through the stack. If it hit another plane, the falling debris would probably take out other planes. The next few minutes were very tense, to say the least, until we were certain the plane and crew had passed through the stack without hitting anyone. As usual, Hump flying was a few hours of tense flying, interspersed with moments of stark terror.

One day I was given a load of pipe to take to Chengtu, a new B-29 base being built in North China. They had put bulkheads in the cabin, and the pipe was stacked to within 3 feet of the top of the cabin. We had to crawl on our bellies over the top of the pipe to get to the cockpit. If we had an emergency, there was no way we could have gotten out of the plane. When I got to the field, I saw what mass manpower could do. We were told that there were 50,000 Chinese workers on the field. These lines of workers with baskets of dirt on their heads literally moved a mountain of dirt by hand. Others were breaking rocks into gravel. They would sit next to a large rock, about 2 feet in diameter, hold a metal band about 3 inches wide and 8 inches across, and break the larger rocks into a size that would fit into the band. They would then peck it with a hammer until it was broken up into gravel. After that, they would put it into a basket about 6 inches deep and 18 inches across. When it was full, they would stand up, coil a cloth on top of their head, put the basket on their head, and carry it to where the other workers were building the runway. They would empty it, return to their workplace, and start all over. I was later told that a rock crusher was flown into the base, which put over 2,000 workers out of work.

North of Yunnanyi was the Likyang River Gorge, where the Yangtze River ran between two mountain peaks. The mountains were 22,000 feet high, and the river ran between them at 7,000 feet above sea level. The gorge was very narrow and caused a wind tunnel effect. Other pilots had told of flying through the gorge just for kicks. They said you wouldn't do it but once. I had to try it, so one day en route to Kunming, I flew north to the gorge. I let down to about 2,000 feet above the river before it got to the gorge, and I entered the gorge. Going through the gorge, I experienced very rapid acceleration. The wind tends to blow you through the gorge, like shooting rapids in a kayak. The wind screams around the plane like a banshee. When I came out the other end, I had to agree, you only did it once.

The ferry pilots who were bringing planes from the States to use on the Hump were bringing over cases of cigarettes and selling them to us for 50 cents a carton. We were taking them to China and selling them to the Chinese. We would go out at night at Kunming and see a cigarette glowing in the dark, and it would be a Chinese wanting to buy cigarettes. We would show him the cartons we had, and he would pull a roll of American money out of his pocket and give us $25,00 a carton for the cigarettes. We were making a killing on the cigarettes, but all good things must come to an end. Intelligence found out that the Chinese were buying the cigarettes for the Japanese, and they were using the American money that the Japanese had taken in the islands they had captured. We were told to stop selling cigarettes to the Chinese. The American Government had written off the money lost in the Islands, and printed more to replace it. By brining the money back into circulation, we were causing problems.

## TOKYO ROSE

The Japanese had a radio program broadcasting propaganda. The announcer was a woman called Tokyo Rose. She would tell us that we were losing the war, and should go home. Also, that our girl friends and wives were going out with other men, while we were fighting the war, and other things like that. She would play the latest songs from the States and good music. I don't know why the Government was so upset about her. No one paid any attention to the propaganda, but we did like the music. Whenever I checked out a new pilot on the Hump, I always told them to tune in Tokyo Rose, and listen to the music. If they did, they wouldn't be so tense while flying the Hump. I thought her program was great.

72

*Dressed for the Hump – After they FINALLY got warm clothing and boots. 1944.*

# Chapter XVII

# A STRANGE WAR

Japan was at war with China, yet China was getting fresh seafood in the interior daily. The Japanese would run the trains from the coastal cities to the end of the area they controlled and turn them over to the Chinese, who would run them into the interior. They would return the trains to the Japanese, who would take them back to the coast. I guess you could say that the Japanese and Chinese were friendly enemies.

We hauled thousands of tons of war supplies to China, and the Chinese were putting most of it in caves, saving it for the coming revolution. If the process hadn't been so serious, it would have been comical. We would take trucks and jeeps to China, and when the Chinese unloaded them, they would beat on them with hammers to get rid of the "evil spirits." When they drove away, the vehicles looked like wrecks.

The American government made Generalissimo Chiang Kai-shek a multi-millionaire, but then we made millionaires of all the "top brass" in every country we went to. Look at what we did for the Marcos in the Philippines. One day, I was given a very unusual trip. I was alerted late one evening for a night flight to China. When I got to operations, I was given a sealed envelope and told it contained my instructions. I was not to open it until I had passed the point of no return and was committed to going on to China. I was told that my destination was Kunming. I checked the weather, filed a flight plan, and was told that a jeep would take my crew and me to the plane. When we got to the revetment where the plane was, we were stopped by a jeep with a 50-caliber machine gun mounted on it. There was another jeep like it parked under the wing of the plane. I realized that this was not an ordinary flight. The guard in the first jeep said for the pilot to come forward and identify himself. I did and was told to put my ID on the ground and to back away. The guard checked my ID and said for me to have my crew come forward, one at a time, and identify them. After that, I was told to do my exterior inspection, get into the plane, and close the door. The cargo was metal boxes covered with burlap. There were no markings on the boxes. We ran the checklist and started the engines, and I called the tower for permission to taxi. As I taxied out of the revetment, a jeep on each side of the plane trailed me. I got a clearance to take off, and the jeeps followed me as long as they could keep up. I climbed up to cruising altitude and headed for China. They say, "Curiosity killed the cat," and I was dying to find out what was so valuable in those boxes that they would guard them with two jeeps with mounted 50's. When I knew we were going on to China, I opened the envelope. The manifest showed that I had 8,500 pounds of U.S. currency on board. The instructions said when we were thirty minutes out, we were to call the Kunming Tower and give them a code word. I complied, and when I was taxiing in, I was being followed by a tank and two jeeps with 50's. The Chinese put the boxes on a truck, which was followed by the tank as it departed for parts unknown. There had to be millions of dollars in the boxes, destined for Chiang Kai-shek. It was just a part of the millions he bilked from the American government.

Madam Chiang would go to Washington to Congress, and when she walked down the aisle, the Congressmen would fall out of their seats and give her anything she wanted. On one such trip, she informed them that we were not hauling enough supplies to China, even though they were not using all the supplies we were taking them. We were taking hundreds of tons of arms and ammunition to China, yet the Chinese guards at the bases were using homemade ammunition. I saw bullets in their cartridge belts that were not all the same length.

Madam Chiang told Congress that we needed larger planes. Someone said that the Curtiss Wright Company had a new plane, the C-46, which Congress decided to buy and send to India so we could haul more supplies to Chiang Kai-shek and the Madam. We were having a hard time getting PX supplies in India. I was buying toothpaste in Calcutta on the black market because we didn't have any in the small PX at the base, yet we were hauling PX supplies, including Kotex, to Madam Chiang. We risked our necks for that!

# Chapter XVIII

# ROOMMATE DOWN

On November 16, 1943, they told me my roommate Rosco ("Smith") was down on the Hump. He was the second roommate to go down. It was a night flight, and they had a bearing on him when he bailed out. He was over the Naga Hills, which was headhunter country.

That night, I wrote my mother, telling her that Smitty had gone down. At that time, a mail censorship was in effect. All enlisted mail had to be censored by an officer. Officer mail could be censored by the officer, but officer mail was spot-censored by a Censor Officer. The letter I wrote my mother was censored by the Censor Officer, and they took away my censor privilege for three months. They said it disclosed possible death or injury to personnel. Shortly before Smitty went down, the famous war correspondent Eric Severeid was on a plane that went down on the Hump. There was a big write-up about him and the incident in the *CBI Roundup*. It said that there had been twenty-two people on the plane, and all but three had gotten down safely. I heard that they were dropping them fried chicken and ice cream. So much for fair censorship.

I had to drop Smitty survival rations, but then he wasn't a famous war correspondent.

The next morning, after Smitty went down, I took a plane and went to look for him. That afternoon, I found him and his crew. He had found a clearing and spread out their parachutes so they could be seen. I dropped them food, a first-aid kit, and a Box Brownie camera with some film. I included a note telling them to stay there, and I would return the next morning with shoes and a map showing them which way to travel.

The next morning when I made the drop to them, I saw natives with them. I decided if they hadn't been harmed during the night, they were probably safe. They went with the natives to their village. I kept dropping them supplies and items like salt and costume jewelry to give to the natives to help make friends. They were out there over a month. They finally were able to clear a strip of land along a river near the village. It was long enough for a small L-5 plane to land, and they were flown out one at a time.

As I said, they were in headhunter country, and Smitty used the camera I dropped him to take some pictures of fresh heads that had just been taken. Later, unfortunately, while my album was on display with other pilots' memorabilia at a Hump Pilots Convention, the pictures of the heads were stolen out of the album.

Smitty also brought me two headhunter knives, which were 2 feet long and 3 inches wide at the tip, with a bamboo handle. The tip was about 1/4-inch thick and heavy. The end was broken off square. I was told that the reason that the tip was broken off square was that the British government, which ruled Burma at

*Over the "First Ridge," returning to India from the Hump.*

*Sunrise on the Hump.*

that time, had decreed that all knives like the ones the headhunters used had to have square-tipped blades. The natives lived in bamboo huts that were about 5 feet above the ground, and they slept on mats on the split bamboo floor. It seems that they took care of grudges by going over to their enemy's hut at night, slipping under the hut, sticking their knife up through the bamboo floor, and impaling their enemy on the blade. The British rule was an effort to prevent the killing.

I was told that before a headhunter boy was considered a man, he had to use a headhunter knife to cut a goat's head off with one blow. After I returned to the States, I decided to check that one out. We were going to kill two goats for a barbecue, and one of the two goats was tied to a tree, so I took a headhunter's knife and struck the goat on the neck behind the head. I did it the same way the headhunters had, and the goat's head dropped to the ground, with the goat still standing. It worked.

The natives carried the knives in what was called a "Dah board," a board about 12" long and about 6" wide. Across the top of the board and about 1/2" in from either side and the bottom, the board was hollowed out about 1/2" deep. There was a strip of woven reeds across the board, making a slot to keep the knife in. The board had a wide belt on each side to go around the man's waist and was worn on the back, with the handle sticking up between the shoulders. The natives would jump up, pull the knife out of the Dah board with one hand, and as they brought it down, they would grab it with the other hand and strike whatever they were trying to kill. The weight of the end of the knife was such that a tremendous leverage was produced.

In December of 1943, a head village completely wiped out another village, taking over 100 heads. They killed every man, woman, and child in the village. I heard that the head village had done it because the other village had failed to pay their taxes. (Pigs, chickens, cattle, etc.) After that, whenever Sector Intelligence got word that there was going to be a raid, they would send fighter planes to circle the village about to be raided, to scare off the attackers

The co-pilot told me this story on Smitty. He said one day Smitty indicated to the Head man in the village, that he would like to have a headhunter's knife. The man gave Smitty a very large, very nice headhunters knife. Smitty ran his thumb down the blade, and indicated to the man that he didn't think the knife was sharp. The man took the knife and a stone they used for sharpening, and sharpened the knife more. He handed it back to Smithy, who again checked the blade with his thumb, indicating he didn't think it was sharp. The man took the knife back, went over to a small tree, and cut a limb about four inches in diameter, and about six feet long. He stuck the limb in the ground, and put the knife in the Dah board on his back. He danced around the pole, jumped up, pulling the knife out, and came down, slicing the pole in half, without knocking the bottom half over. Smitty took the knife and departed. He decided it was sharp enough, after all.

A British woman missionary had gone into the Naga Headhunter country to try to convert the Naga Headhunters. The headhunters treated her as their Queen.

The British Government frowned on headhunting. However, when the Japanese moved into Burma, they saw a chance to use the headhunters to help them in their fight against the Japanese. They sent a British officer to contact the woman and tell her to tell the headhunters they could take all the Jap heads they wanted. After the war, I heard that the officer went back to the village where the missionary lived, and married her.

# Chapter XIX

# POTPOURRI

On one of my trips back from China, I observed a beautiful phenomenon. Before daylight, I took off from Kunming and climbed out through a solid deck of clouds, topping out at 20,000 feet just at daybreak. The top of the overcast was flat as far as the eye could see in every direction. I was flying almost due west, and as I topped out over the cloud deck, the big, yellow full moon was sitting on the cloud deck ahead of us. On the cloud deck were little tufts of clouds which started tinting pink. I swung the plane around, and the bright orange sun was sitting on the edge of the clouds in the East—an impressive and beautiful sight. It looked as if a giant had rolled the sun to the east and the moon to the west, and they had rolled a few hundred miles out and stopped. The odds of witnessing something like that must be great.

I observed another phenomenon while I was stationed at Tezpur. We were having a very heavy thunderstorm, and all at once, sheets of ice started falling all around the area. I was used to seeing hail, but I had never seen or heard of sheets of ice falling from the sky. The weatherman said this phenomenon was probably caused by a tremendous amount of water in the clouds and a sheer wind of high velocity, which blew it out in sheets that then froze and fell to earth. If one of the sheets of ice hit you on the back of the neck, it could decapitate you.

During the war, we had what was called "Short Snorter" bills. We would get a new bill of currency in every country we went to and have someone (preferably a wheel) in that country sign their name on the bill. We would then attach it to the last bill in the roll of bills. I had gotten Chiang Kai-shek, Claire Chennault, Lord Louis Mountbatton, and others to sign bills for me. The bigger the roll, the greater the prestige. I had a large roll, and someone stole it.

On a trip back from China, I had a load of Chinese military on board and thought it would be a good time to get some of them to sign a new Chinese bill I had, for my "Short Snorter" roll. I told the copilot to fly for awhile, and I went into the cabin, took the bill and a pen, and with a combination of sign language and Pidgin English, I indicated to one of the Chinese that I wanted him to sign the bill. He took the pen, put his "hieroglyphics" on the bill, handed it back to me, looked at me, and—in perfect English—asked, "Where are you from?"

I responded, "In India or my country?"

He said, "In your country."

I slowly said, "I'm from a state called Texas, in America."

He smiled and said, "I like Texas. I graduated from Texas A&M. See that man over there? He is a VMI graduate." He pointed out others who had graduated from some of the top military schools in the States. They were all officers, and I thought that I had a load of ordinary Chinese soldiers. I had used sign language in trying to communicate with them, and I felt like crawling in a hole. It's a small world.

On one trip over the Hump, the cargo was for a field at Yangkai. It had been raining for three days before I arrived. I landed and turned off onto a taxi strip, which was blocked by a B-24 bomber. The "follow me" jeep driver motioned for me to cross the taxi strips. I opened the window and signaled to him that I didn't want to get off the concrete strip. I didn't think the waterlogged ground would support the weight of my plane. The "follow me" driver angrily indicated that I was to follow him across the mud. I muttered a few well-chosen words about his reasoning, as well as his ancestry, and turned off the taxiway toward the jeep. I could feel the wheels sinking in the mud, and all forward movement ceased. I put my head out of the window and said, "I hope to Hell you are satisfied." I got out of the plane to see how bad it was. The main gear wheels were sunk up to the hubs. I told the stupid driver, "You got it in there; you get it out." I told him to have it pulled out backwards—the way it had come in—not to try to take it forward. I took my briefcase and walked to operations. Since it was late, I decided to stay overnight. The next morning early, I went to the flight line to check on my plane. The idiots had tried to take the plane forward, contrary to my instructions to the "follow me" driver. Now the wheels were sunk up to the top of the tires, and the props were almost touching the mud. I thought they were going to have to take the plane apart to get it out.

There was a possibility that if we could get enough power, maybe the plane could still be pulled out backwards. There was an engineering unit on the base, and I asked them for assistance. After surveying the situation, the engineer thought he could pull the plane out backwards. He took two large dump trucks, chained them together, and filled them with rocks. He jacked the wings up, put long 12' x 12' timbers under the wheels, and back up the ditch the wheels had made coming in. They started the trucks and slowly pulled the plane backwards to the taxi strip. If the trucks had not been available, the plane would never have been moved.

One day I was told to take a crew to Gaya, India, to bring a plane back to Mohanbari. While at Gaya, I went to a nearby village called Bud Gaya, where Buddhism had supposedly originated. It was said that Buddha sat under a Banyan tree and got "Supreme Enlightenment." There was a temple on a high hill there, and there must have been 300 steps up to the temple. There were some huge boulders near the temple, and an old man lived under them. He looked as if he hadn't had much to eat, so I gave him a ten-rupee note (about 30¢), and he ran all the way down the hill to the village.

On my fiftieth mission, when I got back to Mohanbari, I buzzed the tower, almost blowing the thatched roof off. When I landed, the tower operator told me that the base commander wanted to see me. When I went into his office, he asked, "What are you trying to do? Become a copilot?"

I said, "No, it was my fiftieth mission, and I was glad to still be alive, so I was celebrating." That morning, one of the pilots had tried to land in dense fog, hit the trees, and spun in, killing all the crew. The "Old Man" just smiled and said, "Be careful. That's all."

On March 16, 1944, I flew my last trip over the Hump. My logbook entry shows, "Topped thunderstorms at 24,100 feet, loaded. Back at 25,000 feet. Lost right engine. Came in on one." It was a fitting climax to my Hump flying.

My first trip was to Yunnanyi on April 10, 1943; my last one was also to Yunnanyi, eleven months later. I had trouble on both flights and a lot of trouble in between. My base commander said I had flown the Hump enough. A few pilots had flown sixty-nine trips, but none seventy. Two or three had tried it so that they would hold the record, but there seemed to be a jinx on the seventieth trip. None survived it. I was ready to quit after sixty-eight. Later, after there were good navigational aids, better maintenance, and no Japanese to keep them from flying south during bad weather, some could pass the seventy mark—and probably did.

## FROM MY LOG BOOK

| Trip | Date | Remarks |
|------|------|---------|
| 1 | 4-10-43 | Blew off course. Nearly crashed into mountains. |
| 2 | 4-12-43 | Drifted off course. Got bearing. Topped at 20,000 feet. |
| 4 | 4-24-43 | Yunnanyi on raid alert one hour. Another after landing. |
| 6 | 5-22-43 | Hit snowstorm. Loaded up with ice. |

*Receiving the first of two Air Medals for flying the Hump, in India.*

| 8 | 6-5-43 | Lost altitude from 21,500 feet to 15,000 feet on instruments. |
| | | Clear ice. Right engine cut out. Landing gear wouldn't come down. Had to bounce it down. |
| 13 | 7-9-43 | Rough instruments. Snow and ice. Prop governor froze. |
| 18 | 7-30-43 | Hard rain. Lost engine. All started to jump. Broke clear. |
| | | Continued and landed on one. |
| — | 8-19-43 | Went to Mohanbari to get C-46. Caught fire on take off. Had to ground-loop to stop after using full brakes and running out of runway. Had 8,000 pounds 60 mm mortar shells on board. Fire out when stopped. |
| | 8-22-43 | Dropped food to Carmac's crew. Couldn't find him. |
| | 8-24-43 | Bad ice. 15,500 feet. Rainstorms. Returned to base. |
| | 8-24-43 p.m. | Went back over. Dropped food, clothes & shoes. No news of Carmac. |
| 29 | 9-9-43 | Snow and clear ice. Hard rain on first ridge. |
| Return | 9-15-43 | Bad weather, snow & ice. Lost both engines in thunderstorm. |
| | | Returned to Mohanbari. |
| 31 | 9-16-43 | Heavy snow & thunderstorms. |
| — | 9-25-43 | All instruments went out, just before landing. |
| 34 | 10-18-43 | Over north route to miss Jap interception. |
| 36 | 10-25-43 | Over extreme north route. On approach, air raid. Went to mountains. Jap's bombed and strafed Tali and Chinese headquarters. |
| 39 | 11-9-43 | Saw Jap Zero. No action. Other plane reported four near there. |
| 41 | 11-16-43 | My roommate Rosco Smith went down tonight. He bailed out. He makes two roommates to go down. |
| 42 | 11-22-43 | Back 18,000. Extreme south drift. Brought back personal effects of Vicks crew that crashed. All killed. |
| 44 | 12-1-43 | Over 19,000 feet. Temp. -10 degrees. No heat. |
| 46 | 12-18-43 | Passengers for Kunming, but Jap's beat me there and put the field out of commission. Landed Chengkung. |
| 48 | 12-21-43 | Two planes lost yesterday on Hump. Brought back two survivors who had walked out of Hump. |
| 49 | 12-25-43 | Started to Yunnanyi, but Jap's beat me there. |
| 58 | 2-2-44 | Wore electric suit for first time. |
| 59 | 2-4-44 | Oil congealed at 14,000 feet Feathered engine. |
| | | Continued and landed on one. |
| 60 | 2-6-44 | Over 19,000 feet. Saw four unidentified fighters. Kunming was on alert. |

| | | |
|---|---|---|
| 61 | 2-9-44 | Took 8,500 pounds U.S. paper money to China. |
| Return | 2-11-44 | Got just past TI, ATC called off flying. Went out again to 21,500 feet. Couldn't get over. Returned again. |
| 62 | 2-14-44 | Kunming had an alert just after I left. Jorhat had one as I approached. |
| 63 | 2-17-44 | Back at 14,000. No clouds. First time in months. |
| 64 | 2-23-44 | Back at 18,500. Lt. Munson killed. Spun in. |
| Local | 2-25-44 | Plane caught fire in air. Got it on the ground, but plane burned up, except right engine & wing. |
| 65 | 2-28-44 | Went to new field in Northern China. 50,000 Chinese Coolies working, building field. |
| 67 | 3-8-44 | Over 18,000. Hump broken. Paulet Goddard due in when I left. Back at 21,000 - ice. |
| | 3-10-44 | Air Freight shuttle. On takeoff at Jorhat, right prop ran away. All elec. controls inop. 3,000 lbs. cargo. Lost right propeller in flight. Didn't know prop was missing until stopped on runway because unable to taxi. |
| 68 | 3-16-44 | Topped thunderstorms at 21,500 feet with load. Back at 25,000 feet. Lost right engine. Came in on one. |

During my last month at Mohanbari, I flew only four days and made my last flight on April 9, 1944. I had been alerted to return to the States and had been given the option of going there for thirty days and returning to the Hump—or rotating to the States permanently. I chose the latter, a decision I would regret later. I had come to India as a flight officer and was returning as a first lieutenant.

# Chapter XX

## GOING HOME

I was scheduled to leave Mohanbari on April 15, 1944. My crew and I were to "deadhead" (ride as passengers) to Agra, where I was to pick up a C-46 and return it to the States. I had been assigned a crew, and my roommate Pop Allman was to be my copilot. He was anxious to go home as his wife had had their first child after he had gone to India. When we got to Agra, Pop and I went to see the plane. I couldn't believe what I saw. The fuselage was sitting there with no engines, and the wings were lying on the ground on either side of the fuselage. The plane was the same one that I had lost the prop on. The crew was putting it back together. It seemed that it had been used for parts, and now they were rebuilding it to send to the States as a "war weary plane." I checked the maintenance forms, and the plane only had 290 hours on the airframe, and it would have all new instruments and two new engines. How stupid can people get? I had just been flying the Hump in worn-out planes with more than 1,000 hours on them, and they were sending this one back to the States as a "war weary" aircraft. It was commonplace to use a plane for parts that had a major problem because of the system's failure to supply adequate parts, but you don't send a rebuilt plane like this one back to the States when it is needed on the Hump.

They told me it would take about a week to get it back together, test-hopped, and ready for us. Pop said he didn't want to wait another week. The last letter he got from his wife said that the baby was sick. He said he was going to leave on the next "Fireball," the name they gave a C-54 that shuttled between the States and Karachi, bringing new pilots for the Hump and taking pilots back who had put in their time on the Hump. I tried to tell Pop that there would be a backlog of passengers in Karachi waiting for the Fireball. He would get out as fast or even faster on our plane. Besides, being a crewmember was a lot better than being a passenger. I couldn't convince him, and he left the next morning for Karachi. The Director of Operations assigned me a new copilot, Captain Dean Djaden, a pilot I knew from Mohanbari who was also rotating back to the States. The navigator that they assigned me was the navigator who had been briefing pilots for flights between Agra and the States.

He was as good as you could get. The crew chief was a crew chief on C-46's ever since they got to India, so he, too, was fully-qualified. I could not have picked a better crew had I done it myself. They got the plane ready sooner than expected, and three days later, we left for Karachi just after lunch.

When we got to Karachi, there was a dust storm blowing in off the Sind Desert. The visibility was almost down to zero. There were some B-29 bombers from the States trying to land. I think they were with the twentieth bomb group. Traffic control told me to hold at a beacon a few miles from the field. One of the B-29's was trying to land. He had made several passes but couldn't get lined up with the runway. He apparently had no procedure for making a poor visibility approach. Finally, I asked traffic control to move him out of there and let me come in, as I was getting low on gas. While flying the Hump, I had developed a timed pattern for poor visibility approaches because we had so much fog and heavy rain, which restricted visibility. They moved him out, and I landed on the first attempt. The B-29 made a couple more passes, and they sent him to a fighter base north of Karachi, where he ran through two buildings as he was landing. Another B-29 was lost that same day when it caught fire on take off.

After checking in with operations, we got a room and took a shower. Afterwards, we went to the club, and Pop was there. He was surprised to see us. I told him that Djaden had replaced him and that I would tell his wife he was on his way home. I wanted to needle him for not waiting on the plane. He said he was scheduled to leave on the Fireball at midnight. Because my plane had been classed as a war weary plane, we were restricted from flying at night or carrying passengers. I planned to leave at first light. When we got to operations before daylight the next morning, the Fireball was still there. They were getting ready to leave. Again, I told Pop I would tell his wife he was coming. When the Fireball taxied out, I was right behind it. It took us four hours to get to Masiri Island, where we refueled. It took six hours and ten minutes to get to Aden, Arabia, where we refueled again. Our next stop was Karhtoum, Egypt. We had flown fifteen hours that day. The Fireball was still there and was to leave at first light. I also told operations I would leave at first light. We both filed for the same field, Madugari, Nigeria. Before we got to Madugari, I heard the Fireball call Control and change their destination to Kano, Nigeria, which was about another two hours' flying time. I checked our gas, made a few mental calculations, picked up the mike, and changed our flight plan to Kano. I looked at Djaden, and he was shaking his head. I said, "We'll make it." Then the unexpected happened. The Fireball called in for landing instructions and was told the field was closed and to return to Madugari. I picked up the mike and asked Kano why their field was closed. They informed me that they were having a locust migration, that locusts were a hazard to flight, and that I should return to Madugari. I told them I was low on fuel and had to land at Kano. The Fireball was also landing at Kano. I was flying at 5,000 feet, and I had noticed what I thought were small birds. They were locusts, and the closer we got to Kano, the thicker they got. Soon, there were dark clouds of them. They were from the ground up to more than 5,000 feet. We could smell them burning in the engines, and so many had smashed against the fuselage, wings, and tail that we were losing airspeed. I couldn't see forward through the windshield. I found the field by looking out the side window. I flew the poor visibility pattern and landed.

After parking, I checked the outside of the plane. There was a thick layer of locusts splattered on the nose, leading edges of the wings, tail, and engines. We dug handfuls of dead locusts out of the oil cooler air scoops. I knew that a lot had also gone in the carburetor air scoops. Later the Fireball pilot told me that his copilot had talked him down, by looking through a small place on his windshield that was not covered with locusts. When we got to operations, the operations officer told us to look outside. There wasn't a living plant to be seen. The locusts had devoured everything in their path before moving on.

The next morning, when I went to operations to file a clearance, they didn't want to let me go. When a new engine is put on a plane, there is a special inspection that is supposed to be performed after the first twenty-five hours of flight. I had passed the twenty-five hours before I got to Kano. I told the maintenance officer that his people couldn't perform the kind of maintenance that my plane needed. I was sure that there were parts of locusts in the carburetor induction system, which meant that the whole induction system would have to be dismantled. I wanted to go on to Accra on the coast of Africa, where they had depot maintenance facilities. The next flight after Accra was 1,300 miles over the ocean to Ascension Island, and I didn't want any locusts on the flight. The maintenance officer said he would let me go if I would sign the exceptional release in the flight log. I said that would be fine with me. We started up, and on the way to the runway, the engines were running rough. I knew it was because of the locusts the engines had ingested. On the takeoff, both engines were popping as they swallowed locusts, but we made it to Accra with no problems. We stayed there a couple of extra days, while they scrubbed all of the locusts off the fuselage and stripped and cleaned the induction system. We enjoyed those two days on the beach, learning to bodysurf.

On April 26, we departed Accra for Ascension Island, a trip that took over seven hours. On Ascension Island, we came in over a bluff, and the runway fell away from us. The landing was a little tricky. We refueled and took off for Natal, Brazil, with that leg taking us seven hours and forty-five minutes. About two hours out of Natal, we lost our main hydraulic pump (the C-46 has hydraulic-boosted controls). Losing the hydraulic system meant we had to exert extra pressure to move the controls. Even though Djaden and I took time about flying, we were both worn out by the time we got to Natal. Another problem was there was no hydraulic pressure to lower the landing gear or flaps, which had to be hand-pumped down. The brake accumulator had enough pressure remaining for three to five applications of the brakes. Dean pumped the gear and flaps down, and we landed without incident. The tower told me to follow the "follow me" jeep. Everything was fine until he went down between two rows of parked planes and my brakes went out. We headed for a plane on my left. I gave the left engine power, and we swung over toward a plane on my right. A plane had pulled out of the line on my left, so I added power to the right engine, pulled into the hole and ground-looped to a stop without hitting anything. I had developed a problem on the Hump of having my throat muscles restrict my breathing when I was stressed. I would start coughing, cough the air out of my lungs, and be unable to get air back in. At that point, I would start to gag. When we got stopped, they wouldn't let us out of the plane until after they had sprayed the inside for insect control. By the time they opened the door for us to get out, I was on the floor, trying to breathe. I got out of the plane, went over, and sat down under the wing, still trying to stop coughing. At that point, a jeep came roaring up, and a little sawed-off major jumped out and asked us who the pilot was. One of the crew told him I was. He came over to me and asked, "What were you doing, taxiing without brakes?"

I responded, "I had brakes."

He said, "You couldn't have. You lost your hydraulic system."

Between coughs, I said, "The tower told me to follow the jeep."

He said, "Who was flying the plane? You or the tower?"

I stood up, looked down at him, and said, "By God, I was." I turned and walked away because I knew if he opened his mouth again, I was going to put a fist in it.

The base at Natal was a support base for planes going to the Hump, so they had plenty of parts to repair my hydraulic system. We were able to depart the next morning for Belem, Brazil. We refueled and continued on to Trinidad Island. The flight to Belem took five hours and five minutes; the flight to Trinidad took another six hours and five minutes. We had put in a full day and were looking forward to spending the night at Trinidad. We had heard the song "Rum and Coca-Cola," which said you were guaranteed a real fine time. When we checked into operations, we were handed a pamphlet, on the front of which was drawn a picture of a palm tree with the moon and a girl dancing under the palm, as in the song. Under the picture was written, "Soft moonlight spilling through the palm trees, barefooted maidens dancing in the moonlight, Romantic music filling the air. A beautiful place was Trinidad. For Trinidad as you will find it, turn the page." On the next page was a drawing of a man looking out of a window with bars on it. Under it was the following: "You're restricted to the base." What a letdown.

The next morning, we departed for Puerto Rico. We had picked up lunches at Trinidad, and I planned to refuel at Puerto Rico and to continue on to Miami. We landed at Puerto Rico, and while they were refueling, I went into operations to file a clearance to Miami. A young second lieutenant was the clearance officer. When I handed him the clearance, he said he could not clear me to Miami. When I asked why, he said that there was a bad front between Miami and Puerto Rico, and Miami was not clearing anything to Puerto Rico, so they were not clearing anything to Miami. When we were in Accra, they asked me if I would take a sergeant with me to the States. His mother was dying, and he had an emergency leave confirmed by the Red Cross, and we were the only plane going to the States. I had agreed to take him, although I was not supposed to carry passengers. I told the lieutenant that I was trying to get the sergeant to the States as quickly as possible. Furthermore, I had been flying the Hump in what was considered the worst weather in the world, and I thought I could make it through the front. He said he still couldn't clear me. I asked him who could, and he said I could talk to the major down the hall. I wondered if he was another major, like the one in Natal. When I walked into his office, there sat an ex-airline captain who had gray hair and a pleasant smile. He stood up, put out his hand, and said, "Can I help you, Lieutenant?" I told him what I had told the clearance officer.

He said, "That's pretty bad weather out there. It came through Miami with big hail, and they aren't clearing anything." I told him it was urgent that I get the sergeant to the States as soon as possible. He asked me if we had been to lunch, and I told him we had not. He said, "Why don't you go eat and come back? We will have better information on the front by then." We ate, and when we got back to operations, the major said there had been no change in the weather report. Finally, he said, "I'll clear you at pilot's discretion. You must sign the clearance." I told him that was good enough for me, thanked him, and went back to operations. I told the crew to go get the plane ready while I filed a clearance. When I gave the clearance officer the clearance, he asked to see my instrument card, which authorized a pilot to fly in bad weather. Even though I had been instructing instrument flying at Homestead Air Base before going to India, I had never been given an instrument card, I told the lieutenant that I didn't have one, but my copilot had one, and if we got into weather, I would let him fly. That seemed to satisfy him. On my way to the plane, I had a hard time imagining there was bad weather anywhere, as it was a beautiful day, the sun was shining, and the birds were singing.

We took off and headed for Miami. When we were about an hour out, we could see a dark line on the horizon. I noticed that the waves were starting to get whitecaps below us. I told Djaden we could fly over the weather. I had been up to 30,000 feet in a 46 empty. I pushed the throttles up to climb power and started to climb. The closer we got to the front, the higher the clouds looked and the higher the waves below us looked. I realized we could never top the clouds. The clouds were a dark-greenish purple color, which meant hail. I asked Djaden if he was thinking what I was thinking. He said, "If you're thinking we would be a fool to fly into that, we're thinking the same." I looked to the right, and it looked the same as far as I could see; I looked to the left, and it looked as if it was better in the distance. I decided to parallel the front to the left to try to find an area where we could get through. I didn't want to go back to Puerto Rico. We were too near home.

We flew about a half an hour and were almost to Havana, Cuba. It didn't look too bad, and there were no indications of hail. I told Djaden it was now or never. I let down to just above the waves, zeroed the altimeter, climbed up to 2,000 feet, and turned into the front. Just after entering the clouds, we ran into a tremendous rain, a deluge, but no hail. After about an hour, we came out the other side in the clear. I told the navigator to give me a heading for Miami.

A short time after we headed for Miami, a strange thing happened. All of the compasses quit working. They would just spin and not show any reference to north. We were in some kind of magnetic field that prevented the compasses from working. I had heard of the Bermuda Triangle and the weird things that had happened there, but we were in the wrong area for that. The navigator said to try to hold the same heading, and we had to hit land somewhere along the coast. Eventually, the compasses started working normally, and we were able to get a bearing on Miami. What a relief it was when we were able to see the skyline of Miami.

The airport at Miami still had a 35-mph wind, left over from the frontal passage. Before we got to the parking area, my right brake was smoking from holding against the wind. When we went into operations, the operations officer said, "Where did you come from?"

I said, "Puerto Rico."

He said, "You couldn't have."

I handed him my clearance and said, "Read it. It says Puerto Rico."

He said, "Come here. I want to show you something."

He took me to the door and said, "Look at that," pointing at the trees and buildings. The trees were all broken down, and the roofs of the buildings were severely damaged. He said, "When that front came through here, we had hail as big as grapefruit." I told him that where we had hit the front, I could tell it had hail, but we had flown around the worst part. He just shook his head and walked back inside.

My orders were for me to deliver the plane to Brookly AFB in Mobile, Alabama. We stayed in Miami two more days, then departed for Mobile on May 2. I "sold" the plane and thanked the crew, and we split.

# Chapter XXI

## DODGING THE BULLET

They gave me a thirty-day delay in route, and then I was to report to an OROD (Officers Returning from Overseas Duty) unit in Nashville, Tennessee. I think they thought that officers who had been overseas for an extended period of time needed to be rehabilitated. There had been a lot of changes since we had left, such as the rationing of gas, automobile tires, etc. We spent the time there attending briefings on the conditions in the States, as well as the military situation abroad. They also gave us a humorous pamphlet telling us how to conduct ourselves. One page contained a cartoon of a military man blocking a door, and the caption read, "If you get ready to leave someone's house and can't find your coat, don't jump in the doorway and say, 'Don't anyone leave the room until I find my coat.' The hostess probably put it in the hall closet." Another page had a drawing of a commode with the caption, "In the bathroom, you will find that this is not an automatic foot washer." We hadn't regressed that much.

The OROD Unit also gave us our next assignment. The assignment they gave me was the C-46 school at Reno Army Air Base, about 9 miles from Reno, Nevada. They had asked me if I wanted to go back to Homestead to instruct in C-54 transports, and I told them I wanted to stay in a C-46 unit and was happy when they gave me Reno. It was a C-46 transition school for pilots going to the Hump.

On my way to Reno, when I passed Fort Sill, Oklahoma, I saw a young GI standing by the highway, trying to catch a ride. I stopped and picked him up. I was wearing Western clothes, boots, and a cowboy hat, so he couldn't tell I was an officer. He said he was going home on his first furlough. As I continued down the highway, I caught up with two cars. As I started to go around them, the back car started to go around the front car, and I felt his car brush the side of my car. I went on past the two cars and asked the boy to look out of his window and see if the other car had damaged my car. He said it only knocked the dust off. I picked up my speed again and continued down the highway when the car that had brushed the side of my car swerved around me and slowed down to about 35 mph. I honked and started around him. When I did, he moved over in front of me, blocking the highway. I moved back, and he moved back in front of me. Every time I tried to pass him, he moved over in front of me. I put up with that for about 5 miles but got tired of his playing games with me. Since I had a carbine under my seat, I took it out, put it out the window where he could see it, honked, and motioned for him to move over. I knew he was watching me in his mirror, and when he saw the rifle, he got over and stayed over. When I went around him, I looked over and saw he was an army colonel in uniform. I picked up speed again and continued down the highway.

He suddenly passed me at high speed. I told the young soldier that he was probably heading for the police. I didn't want to get trapped between the Oklahoma police and an irate colonel whom I had just moved over with a rifle, so I told the boy to get the map out of the glove compartment and find a road that would get us off the highway. He said there was one just past the next town. I was staying close to the colonel, and when he got to the next town, he stopped at a convenience store, jumped out, and ran in. I knew he was going to call the Oklahoma City police, who were about 30 miles down the road, so I got to the road going south, turned on it, and put the pedal to the metal. I wanted to be out of sight when the colonel came by.

The road went about 20 miles south, turned east, and went to Shawnee, where I was headed anyway. I told the soldier he would have to take a bus to Oklahoma City. When I got to Shawnee, I stopped in a gas station and asked where the police and bus stations were. I took the boy to the bus station and then went to the police station, where I asked to see the chief of police, then told him that I had put the rifle out the window so he could see it and that I hadn't pointed it at him at any time. I explained that I was a first lieutenant in the Air Corps, had just returned from fourteen months overseas, and didn't want to get involved with an army colonel. I told him that I was going to stay at the hotel for a couple of days and if the Oklahoma City police were looking for me, he would know where to find me. I wasn't trying to evade the police, only the army colonel. He laughed and said he didn't think I needed to worry about it, adding, "Enjoy your stay in Shawnee."

I stayed in Shawnee for four days. I waited until after dark before leaving because I had to go through Oklahoma City and thought it would be safe to go through it after dark, being careful to obey all the traffic laws while doing so. I drove all that night, all the next day, and until one o'clock the next morning. I wanted to put as much distance between the colonel and myself—and the Oklahoma City police—as I could. Halfway between Las Vegas and Reno, I pulled off the highway and went to sleep.

# Chapter XXII

## RENO AND THE AIR BASE

The air base was located about 9 miles north of Reno in the desert in what was called Lemon Valley. The valley was surrounded by low sagebrush-covered hills, a dramatic change from the bases and the jungles in India that I had become accustomed to and loved. Reno was not as barren because the Truckee River, which was fed by Lake Tahoe 6,000 feet up in the Sierra Mountains, ran through the city, turning it green. Also, at that time, Reno was the gambling capital of the United States, which also kept Reno green, with "greenbacks," that is. There were some clubs, like Harold's Club, that were honest, but they were in the minority. It was so bad that the base commander had, posted at the base, a list of clubs that were considered "honest" so all would know which clubs to gamble in if that was what they were going to do. Most of the officers who came to the base for upgrade training had never been around gaming casinos, and some of them went overboard and lost a lot of money. One lieutenant, who got killed in a crash, had lost all of his money. His wife had to borrow money to go home. Sometimes, if I didn't have anything to do and didn't want to go to the mountains or play golf, I would go to Harold's Club and play blackjack, but I always had a limit and willpower. If I were losing, when I reached my limit, I would walk out and go to a movie or return to the base. I never did lose very much. Harold Smith, Jr., who owned Harold's Club, was drafted into the army, so Harold's Club looked after the military. If one of the men in uniform lost quite a bit—more than he could afford—it wasn't uncommon for the club manager to return most of it and ask the man to please leave for the time being. The women who were dealers also helped the servicemen. In dealing twenty-one, they would hold their cards so we could see what they had. That way, we would know whether or not to "take a hit."

Reno was a tough town in the 1940s. The word was that most of the police were ex-cons whom the military wouldn't take. I was aware of one incident that supported this belief. A captain I knew got orders to go overseas. He and his wife were living in Reno with a family who had a daughter the same age as the captain's wife. The captain's wife was eight months' pregnant, and her doctor thought it best that she stayed there until the baby came, so they arranged for her to stay with the family, and the captain shipped out.

One night, the captain's wife and the family's daughter were at home alone, when they saw a man looking in the window. The girl called the police, who came, talked to the women, and left to search the neighborhood. They found a man near there and brought him back to the girl's home. The police asked the women if they could positively identify the man as the one who was at the window. They said they couldn't positively identify him because they had seen his face for just a moment. The policeman told the captain's wife, "If you can't positively identify him, we can't hold him. After all, you shouldn't complain. There's a man shortage." When word got to the base about the incident, there was a feeling that maybe we should go into Reno and take on the police to see just how tough they really were.

It is accepted that money establishes a person's position in society, as evidenced in a Sunday issue of one of the San Francisco papers while I was at the base. An article about the socially-prominent families in Reno stated that, in the old days, a housing area across the river was a red-light district called "The Bull Pens." The miners would come into town with their gold, wind up at The Bull Pens, fall in love with one of the women, go back to prospecting, strike it rich, marry her, and settle down in Reno. His riches would make them socially-prominent. The article mentioned that in several of the prominent families in Reno, the women came from The Bull Pens.

When I checked in at the base, I was assigned the duty of flight commander. The base was a transition school for pilots who were already qualified pilots but had never flown the C-46. They were to qualify in the C-46 before going to the Hump. Because of the amount of time I had had in the C-46 (I probably had more time in the C-46 than anyone else in the Air Corps), I was made a flight commander, even though I was only a First Lieutenant. Before the student pilots were considered qualified in the C-46, I gave them their final check ride—and the instructors in my flight their proficiency checks. I also filled in as a clearance officer in operations and as a control tower officer. An officer was required to be in the tower when the field was open to flying. I had a lot of free time and played a lot of golf in Reno.

# Chapter XXIII

## THE UNEXPLAINABLE

One day, when I was the tower officer, a bizarre incident occurred. The tower got a call from an incoming plane requesting landing instructions. They had not received a flight plan on a plane en route to the base. However, that was not unusual. The plane made a normal request for landing instructions, so the tower operator cleared the plane to land, but what the pilot of the plane had failed to say was that he was running out of gas and that he had shut down one engine because of no fuel. The plane was an early model B-24, and each engine operated off a separate tank. By the time he reached the field, he had lost another engine. He still had one on each wing that was running, but his troubles were not over. He overshot the runway and had to go around. He lost the third engine and landed with the gear down, about a half-mile from the nearest runway, in the sagebrush. When I saw him overshoot the runway, I didn't think he could make it around the pattern. I flipped on the crash alarm switch and headed down the ladder to my jeep. I had to work my way through the sage and around or through washes to get to the plane. Just as I got to the plane, a gray-haired general climbed down the ladder, dusted himself off, and asked, "Lieutenant, do you have quarters on this base?"

I said, "Yes, Sir."

He said, "Take me to operations. Have this plane towed in, and service it. I will take off at eight in the morning."

I couldn't believe how calm he was. By all rights, the plane should have been a smoking pile of metal. He was at least a half-mile from the nearest paved surface, and he had landed with his gear down. They took a bulldozer to tow the plane in and had to fill in a couple of washes and blade a path through the sage before they could move it. When they got the plane to the ramp, Captain Peyton Walmsley, the flying safety officer, went out and checked the plane for the accident report. He had to make it an incident report. All that he could put in the report was that the plane had made a landing beyond the limits of the runway. There wasn't even a cut on a tire. General Olds was the pilot, and the next morning, he took off for Sacramento, California, just as he said he would. How do you explain it?

Some things just can't be explained. One clear day I was flying from Reno to Fresno, across the Sierra Mountains, enjoying the beautiful scenery. All of a sudden, there was no pressure on any of the control surfaces. My first thought was that all of the control cables had broken. Of course, I immediately realized that that was impossible. Then, to make matters worse, the plane made a 180-degree flat turn, without dropping a wing. I couldn't imagine what was happening. The fact that there was no pressure on any control surface would indicate that I had flown into a vacuum. However, if that was the case, what had turned the plane 180 degrees? A "wind sheer" would have turned the plane, but it would have also exerted pressure on the vertical surfaces. I have asked many meteorologists if they could explain what had happened, and I have yet to get a satisfactory explanation.

Not long after I got to the base, I was informed that I was to go to Washington. It seemed that someone who had lost a loved one on the Hump had heard that the Curtiss Wright Company, which made the C-46, was advertising that the C-46 was tested on the Hump. They had asked their Congressman why it had not been tested before it was sent to the Hump. Congress had called for an investigation of the matter and wanted a pilot, an operations officer, and a maintenance officer from the Hump to testify. Since I had the most time in the C-46 on the Hump, I was selected as the pilot.

A major was scheduled to fly me to Washington. There was a colonel on the base who wanted to go to Cincinnati, had heard about the trip to Washington, and asked the major to take him to Cincinnati first. The major saw the chance to make a few brownie points, so he agreed, even though his mission was to take me to Washington. When we got to Cincinnati, it was after dark, and the field was covered with heavy ground fog. Instead of changing his flight plan and continuing on to Washington, the major decided to land. He dragged the field two times in the fog but was unable to land. On the third attempt, he was able to get the plane on the runway. When we taxied into the line, the fog was so thick that the lights in the hangar were dim. The next morning, the field was still closed due to the fog. There was a range of hills paralleling the runway, and when the fog lifted and I saw the hills and realized that we had dragged the field two times in the fog and the major wasn't familiar with the field, I couldn't believe how lucky we were to be alive.

It was mid-morning before we could take off. The hearing was to start at eight o'clock that morning. The major had an instrument clearance to Washington. However, in the clouds there was a hole known as a "Sucker Hole," and he let down through it. The trouble was, he was west of the mountains that lay between him and Washington. He saw a canyon and went up it until it closed in with clouds. He had to turn on a wing tip to go back down the canyon. He paralleled the mountains north until he saw another canyon and started to go up it. At that point, I was prepared to take over the plane, even if I had to use the fire extinguisher on the major. He was going to get us all killed. (A lot of planes and crews have been lost doing the same thing.) I saw a river that would take us through the mountains to Washington, so I tapped the major on the shoulder and said, "Major, do you see that river? You get this plane over there and follow that river to Washington. Don't try to go up another canyon." He did, and we got to Bolling Field without incident.

I had a staff car take me to the hearings. I was told that I was scheduled to be the first witness. Thanks to the major and his "brownnosing" and stupidity, I was too late for the hearings. I would like to have been able to testify as I had more experience in the early model C-46 than anyone else.

As I said earlier, when I was due to return to the States from the Hump, I was given the option of taking a thirty-day leave and then returning to the Hump *or* making a permanent change of station. I chose the latter and said at the time that I would probably regret my choice. I had not been in the States two months before I knew I had made the wrong choice. The States had changed, what with rationing, etc. I also missed the excitement of the Hump and my trips into the jungle. Kipling said, "If you have ever the heard the jungle calling, you will always heed that call," so when word came down that they were looking for a few men to go to India to be a search and rescue unit, I immediately applied for the position and was accepted. I had not been able to shoot an elephant when I was there before because I didn't have a big enough gun. I wanted to take a large caliber rifle back with me, not only to hunt elephants but also to feel safer in the jungle with a bigger gun. I started looking for a gun and found that Winchester Arms Company had one H&H Magnum 375-caliber rifle in stock and said they could give me delivery on it in ten days. The price was OPA ceiling price of $79.80. I told them to send it COD. The rifle came, but my dream was broken because my orders had been cancelled. The gun was too big to hunt with in the States. I later sold it for $100.

Shortly after I got to the base, an awards ceremony was held. I was presented my second Air Medal, as well as the Distinguished Flying Cross for my flying on the Hump. Colonel Gur, the base commander, pinned the medals on me. I was very proud.

While I was at the base, something happened that would affect the rest of my career in the military. One day, Major Harralson, my boss, asked me if I had enough time in grade as a first lieutenant. I told him I thought I needed three or four more months to be eligible. He said to let him know when I was eligible, and he would submit my name for promotion. I checked on the time required, and when I had enough time, I told him. He said he would take care of it, but when the promotion list came out, my name was conspicuous by its absence. I asked the major about it, and he said, "Didn't I put your name in? I'll do that right away." It was too late, though. The war had wound down, and they were reducing the size of the military, so they stopped the promotion cycles. Since I was an officer in the reserves, I could hold one rank on active duty and a different one in the reserves. I had made captain in the reserves and would not serve in that grade until I got off active duty and returned to the reserves. Two months before I was eligible for promotion to the rank of major in the reserves, I was recalled to active duty for the Korean War buildup. The regulations required one to be on active duty one year before being eligible for promotion. The Korean War didn't last very long, and again they were having a RIF (Reduction In Force). I had decided that if they were going to keep interrupting my life, I would stay for the twenty years required for retirement. As it happened, I was a captain for sixteen years before making major, thanks to Major Harralson. It also prevented me from getting higher than lieutenant colonel before retiring.

In the latter part of 1945, word came from Washington that they did not need

*Receivng the Distinguished Flying Cross, and second Air Medal for my missions on the Hump, presented by Colonel Gur, Base Commander, Reno.*

# THE UNITED STATES OF AMERICA

TO ALL WHO SHALL SEE THESE PRESENTS, GREETING:

THIS IS TO CERTIFY THAT
THE PRESIDENT OF THE UNITED STATES OF AMERICA
AUTHORIZED BY ACT OF CONGRESS JULY 2, 1926
HAS AWARDED

## THE DISTINGUISHED FLYING CROSS

TO
Captain Alan W. Saunders, AO 1896702
United States Air Force
FOR
EXTRAORDINARY ACHIEVEMENT
WHILE PARTICIPATING IN AERIAL FLIGHT
Asiatic-Pacific Theater of Operations, 18 April 1943 – 30 October 1943
GIVEN UNDER MY HAND IN THE CITY OF WASHINGTON
THIS    28th    DAY OF September    1949

*Distinguished Flying Cross award.*

any more pilots for the Hump, so they were going to close the base. We had to set up a plan for ferrying the planes to storage bases. Some would go to Tucson, Arizona, and the others would go to Walnut Ridge, Arkansas. I would lead the flights of three to five planes and bring the crews back to Reno. On the Arkansas flights, I would schedule them to spend the night at Midland, Texas, and I would fly to Goodfellow AFB Texas. My parents lived 60 miles from Goodfellow, and they would meet me there, and I would spend the night at home. The next morning, I would meet the other planes near Dallas, and we would proceed to Walnut Ridge. I had planned with the other pilots beforehand to do a tactical approach at Walnut Ridge. I would call the tower and ask for permission to do a tactical approach. We would buzz the field at low level, go out, and return in a tactical approach. Upon reaching the field, I would pitch out, and each succeeding plane would follow. It was quite a sight to see a plane as large as a C-46 do a tactical approach. When we buzzed the field, the workers would all go outside and watch. We got a kick out of it also. On the trips to Tucson, we would race to see who could get on the ground first. Sometimes, the traffic pattern would get rather hair-raising. Since they were going to storage, they would send anything that would fly. Some of the planes only had engine instruments that showed manifold pressure and rpm. No flight instruments or gas gauges. Because of the lack of flight instruments, we couldn't fly in clouds, or else we wouldn't have had a visual reference to the horizon. On one trip, the right engine prop rpm ran away, exceeding the limits. I landed on one engine and didn't write up the overspend. If someone bought that plane, they got a lemon.

A base operates on two types of funds: appropriated and non-appropriated. The appropriated funds are provided by Congress and earmarked for specific expenditures. The non-appropriated funds are funds obtained from the base exchanges, commissaries, and clubs and are used for the upkeep of those facilities and for welfare and recreation. When a base closes, regulations state that all non-appropriated funds remaining will be turned in to the regional non-appropriated office. The Officers Club Board of Directors at Reno AFB met and decided they didn't want to turn in their excess funds to the regional office. They would use them for the benefit of the officers. They would have free steak dinners and drinks every Saturday for the members, which didn't use up the funds, so they made free meals on Saturday and Sunday. When that still didn't help, they threw in Wednesdays, which *still* did not help. They suddenly realized why. The Club had about twenty slot machines in the lobby which were set at 60/40, whereas the ones in the clubs in Reno were set at 80/20. The reason they couldn't use the funds was that the members coming to the free meals were putting more money into the machines than the club was spending for the food and drink, and that at 60/40. You can imagine how much money the clubs in Reno were making with their machines set at 80/20.

# Chapter XXIV

# GROUND JOB

My new assignment was at the base at Stockton, California, a Transport Hub, designated MAT (Military Air Transport). The base had three outlying bases that it serviced: one at Denver, one at Tucson, and one at McChord AFB Washington. The base was the forerunner of MATS (Military Air Transport Service). The Airlift concept had its beginning on the Hump. The Stockton Base was the next step, followed by MATS, and then MAC.

For the first time since getting out of flying school, I did not have a flying job. I was assigned as squadron adjutant. I got enough flying time for pay but spent most of my time with administrative duties. One day the director of personnel called me to his office and said he was giving me a new assignment. It seemed that they had been having a problem in the equipment section, and I was to take over that function.

I took an inventory of the equipment on hand, added it to that which had been issued, and compared that to all that had been received. It balanced, so I signed for the account. Later, I was checking the issue cards, and I noticed a card on a pilot who I knew had been transferred to another base. I started checking further and found more that were no longer in Stockton. I instructed the sergeant to call all of the squadrons to see if the men who had active issue cards in our files were there. He made a list of over 100 names of men who were no longer on the base but had not cleared their supply account and turned in their equipment before leaving.

I took the list and went to base personnel. I checked it against the morning report, the official list (which is sent to Washington each day) of personnel present for duty. I found that the Morning Report was in error by about 100 men. When I told the director of personnel that his Morning Report was in error, he was furious. He didn't believe me, saying I had better have proof before making a statement like that. I had done my "homework," so I asked him to call in the clerk with the Morning Report. When the clerk brought in the Morning Report for that day, I asked him where Captain John Smith was. He checked the report and said he was in Squadron A. I told him, "According to Special Order So and So, dated such and such, he was transferred to Denver." I asked him where Lieutenant Jim Jones was. He said he was in Squadron B.

I said, "According to Special Order Number So and So, dated such and such, he was discharged from the service." I looked at the director of personnel, and he was livid. He told the clerk he wanted all section heads in his office on the double. He told me that to clear my account, I had to send a letter to each of the men who had not turned in their equipment, advising them that to have the equipment was misappropriation—and embezzlement of the federal government. If they returned the items, no charges would be filed.

We sent out letters and started getting items from all over the country. A man in Florida returned a parka. I don't know what he would do with a parka in Florida. A lot of the letters were returned "unclaimed." I had to keep all of them to satisfy the director of personnel.

I thought a lot of property was being taken off the base, so I got permission from the base commander to have a shakedown at the gate. We picked out a date, and just before quitting time, we set up tables along the exit, stopped every car, and checked the glove compartment, trunk, and inside of the car. If they had on an issued watch, we would take it, give them a receipt, and tell them they could get it back if they produced a receipt. Although we had recovered a lot of the equipment, we were still short.

One day a military policeman (MP) brought a young airman to see me. He had been discharged and was leaving for home. They checked his bag at the gate and found a military blanket. The MP said he was supposed to take the man into Stockton and turn him over to a U.S. Marshall. He could be charged with theft of government property. I couldn't see that happening, since I had records of officers taking more valuable things like watches. I told the MP that I would get a staff car to take the man to Stockton, and he could go with us.

I took the MP into the next room and told him that when we got to the courthouse in Stockton, I wanted him and the driver to take a walk and be gone about ten minutes. I wasn't about to turn that young boy over to the Marshall and ruin his life.

When we got to the courthouse, I asked the MP if he didn't have someplace to go. He said he did, and the driver was going with him. After they left, I winked at the airman and said, "I've got to go down the street. If you are not here when I get back, I won't know where to look." He understood what I meant, and when I got back, he was gone, and we went back to the base.

# Chapter XXV

## ACTIVE RESERVE

In mid-1946, I was released from active duty. I wanted to stay in the reserves, so I went to Kelly AFB in San Antonio and signed up with a medical evacuation unit there. A friend of mine was the operations officer of the unit. I would call him each month, and he would schedule me for a trip. At first, I was flying to the West Coast. Later, I was changed to the East Coast. I was getting more time than most of the active duty pilots. Unless a pilot was a medi-vac pilot, he could only fly four hours a month, just enough to keep up his proficiency and qualify for flying pay. He could not take cross-country trips.

The East Coast trips were from Kelly to Brookly AFB Alabama to Bolling AFB Washington to Westover AFB Massachusetts. We would remain overnight there, and the next morning we would reverse the route and return to Kelly, getting back there at night.

On one trip, we had left Brookly AFB and were flying past New Orleans at night and in bad weather, with occasional thunderstorms. We had seven litter patients and thirteen ambulatory patients on board, along with a nurse and flight attendants. I had told the nurse to be sure everyone was strapped down well because the turbulence in some of the thunderstorms was very heavy.

We got a call from traffic control, who said a tornado had touched down near McComb, Mississippi, and they didn't know which way it was going. Shortly after the call, we were able to tell them which way it had gone as we had flown into the edge of it. All at once, we were all over the sky. The lightning was a continuous blinding flash. I turned all the lights in the cockpit on full bright so the lightning wouldn't blind us so that we wouldn't be able to see our instruments. We were both on the controls, just trying to keep from turning upside-down or spinning in. Large hail was pounding the plane, and it was raining a flood. Fortunately, we got past it without anything serious happening. It was "touch and go" for a while, but we continued on to Kelly without further problems. I was glad when that flight was over.

# Chapter XXVI

## RECALL TO ACTIVE DUTY

In mid-1951, I was notified that I was being recalled to active duty. I decided that if the government was going to interrupt my civilian life, I would stay in until retirement and try to outlast them. I was told that if I stayed in until I had twenty years active duty, I could retire with 50% of my active duty pay, or 2.5% for each year of pay eligibility. Further, I would receive all the entitlements that I had while on active duty. I was to find out (after I retired) that what they had said was a big lie. It mainly concerned medical care. It was okay at first, but eventually they quit providing different care. At first, we were given a complete annual physical, the same as when we were on active duty. Then they required us to make appointments with each separate clinic. Then they would ask if I was active duty. When I said I was retired, I was given a lower priority. Next, they started cutting back on dental care. At first, I got full dental care, crowns, fillings, root canals, etc., but then they stopped everything but fillings. Then they stopped all dental care.

Next, they set up a central appointments office. They made all the appointments for the different clinics. The problem with that was that you had to make an appointment the first of one month to be seen the next month. They were trying to make it so difficult to see a doctor that we would resort to civilian care. It seemed that every time Congress cut back on funds for the military, they would compensate by cutting back on medical care for retired personnel. The reduction of medical care at military hospitals didn't bother the Congressmen— as they were well-taken-care-of at the hospitals at the military installations in Washington.

My first assignment was to a base at Mountain Home, Idaho. It had been deactivated after the war and was being reopened. The buildings were wartime construction of plywood covered with tarpaper. When they started renovating the buildings, I think every sheepherder in Idaho became "expert" plumbers, electricians, carpenters, and roofers. A list of building numbers would be released for bids, and "contractors" would bid on each building by number. Some "contractors" would just take the list and put down an estimate for each building by number without even having gone to the building. The military always numbered every structure on the base. The practice of submitting bids without first checking the buildings was made abundantly clear when one bid was received for several thousand dollars to rehab building number two. As it happened, number two was the number given the flagpole. I was told that the "contractors" covering the floors with roll floor covering demanded and got a contract for twelve-hour days, seven days a week at $10 per hour, with time-and-a-half for time over eight hours a day—and double time on Saturdays and Sundays. The rehab of the base was a crash program, so the government paid whatever the sheepherders wanted. Later, because they realized that the government had been taken, Congress conducted an investigation of the contractors and contracts.

The base at Mountain Home was the most desolate, demoralizing base I had ever seen—and I've seen some bad ones. It was about 5 miles from the small town of Mountain Home and about 60 miles from Boise. There was nothing in between but sagebrush, coyotes, and rattlesnakes. The morale of the men was a big problem. There were several thousand men on the base, and the only recreation was a 400-seat theater, so the enterprising people of Mountain Home started building shacks to rent to married men at ridiculous prices. The base commander finally had a meeting with the city officials and told them if they didn't stop "gouging" the men, he would put the town of Mountain Home "off limits" to all military. That put a stop to the "gouging." Some of the married men with cars rented houses in Boise and commuted to the base each day. The unmarried men were stuck on base.

There were three wings on the base: the Base Wing and the 580th and 581st ARC [Air Resupply and Communications] Wings. It was a cover for the real mission of the wings, which was highly classified. Foreign "agents" were trying to find out what the mission was.

I was assigned as squadron commander of the 580th Installation Squadron. As part of the training, we occasionally conducted "survival" trips up in the mountains. On one such trip, some of the men were looking in some mine shafts dug into a creek bank. In one they found a tobacco can on a ledge, opened it, and saw four current Communist membership cards. The Communists had apparently been using the mine as a meeting place. The men brought the cards to me, and when we returned to the base, I gave them to the OSI (Office of Special Investigation), who were very happy to get them. OSI later told me that one of the cards belonged to a prominent rancher in the area whom no one suspected of anything. The careless mistake of leaving their cards in the mine blew his cover.

I had one of the best first sergeants I had ever known. He was firm but fair, extremely efficient, and very good with the men. Occasionally, I would walk through the barracks in the evenings just to see how the men were. On one such occasion, I saw one of the men with blood on his mouth. I said, "Garland, what happened to you?"

He stammered and said, "I went to the dentist today, and my tooth came out."

After I went back to my office, the first sergeant told me what had really happened. It seemed Garland had been caught cheating in a poker game, and one of the players had hit him in the mouth. The next morning at roll call, I told the men that I didn't know if I would recommend that dentist, as Garland said he had gone to him yesterday, and last night one of his teeth came out in the barracks. A ripple of laughter went through the formation. I said, "The only thing I have to say about that is, if you feel a tooth getting loose in the barracks, go outside. I don't want any more teeth falling out in the barracks." They got the message: No fighting in the barracks.

In some squadrons, the commander wouldn't permit gambling in the barracks. However, there was practically nothing for the men to do when they were off duty, so I permitted gambling as long as it was done peacefully—although I wouldn't permit any outsiders to gamble in our barracks. Some card sharks would go around to different barracks and take the men to a cleaning.

The 580ᵗʰ Wing was getting ready to deploy overseas. At a group staff meeting, we were told that the squadron commanders and the men who were needed to pack the equipment for shipping would go on pre-embarkation leave ahead of the troop movement. Upon return from leave, they would go on a troop train to the port, which was Camp Kilmer, New Jersey. Everyone would get a thirty-day leave before going overseas.

When the men going on the troop train got back from leave, I let the rest of the men go on leave. They were told to report directly to the port at the end of their leave and they had to leave an address where they could be reached, if necessary. I mentioned earlier that the mission had a very high classification. The squadron commanders had been briefed on the mission. However, there had been so much loose talk, it wasn't difficult to figure out the mission of the Wing.

Everything went like clockwork, until a "colored" sergeant got on the "Welcome Traveler" program in Chicago. The MC said, "Hi, Sergeant. Are you traveling far?"

The Sergeant said, "I sure am."

MC: "Where are you going?"

Sergeant: "I'm going to Tripoli."

MC: "What are you going to do there?"

Sergeant: "Well, I'm with this cloak-and-dagger outfit. . . "

And click. They cut him off the air. However, the damage was done. He had hardly stopped talking, when Washington was on the phone to Mountain Home, wanting to know what the hell was going on. Didn't we brief the troops on what they could and couldn't say? A Chicago paper reported that this incident was reminiscent of the saying during World War II, "The slip of the lip can sink a ship." It said, what the sergeant had started to say could have put us in another war.

An emergency meeting of all commanders was held. We were given a briefing sheet that we were to read to all the men. It said that the only thing they could say was, "We are doing peacetime training for a wartime mission."

Due to what had happened in Chicago, Washington cancelled the movement overseas. We were instructed to send telegrams to all the men on pre-embarkation leave, advising them that the movement and their leaves had been cancelled and that they were to return to Mountain Home at once.

A lot of the men hated Mountain Home so much, they just ignored the telegram and stayed home or went somewhere else, anywhere but Mountain Home. When they didn't return, they were listed as "AWOL" (Absent Without Leave), and their names were sent to all military units. When they were picked up by the military police, we would be notified and would have to send two men to bring them back. Shortly, the brig was full, and all we could do was restrict them to the base. Some of them just stayed around long enough to get clean laundry and go. There was a joke going around that so many were leaving, they had to run extra buses to keep them from crowding at the gate. The wing was below 50% strength, and we had to send a report to Washington every day, saying how many had left that day, how many had returned, and how many were still gone. According to regulations, when they were AWOL over 30 days, they were listed as deserters, a classification category that carried a severe penalty if the deserter was ever caught.

When the married men who had their families with them at Mountain Home left for overseas, they moved their families back home, gave up their houses, and (in some cases) sold their cars at a loss. When the movement was cancelled and they returned to Mountain Home, they put in pay vouchers to recover the cost of moving their family's home. The finance officer told them they would not be reimbursed because they hadn't gone anywhere. Some of the married men went AWOL at that point.

Shortly after the move was cancelled, a C-54 from ARC Headquarters in Washington arrived at Mountain Home. On board were a new wing commander and a complete staff. The new commander was Colonel Kane, also known as "Killer Kane," having made this name by leading low-level bombing raids against the oil fields in Europe during World War II.

Colonel Kane called a meeting of the wing staff. At the meeting, he started with the wing commander, relieving him of his command. He then went through the entire staff, saying, "You're through. Report to Washington for further orders. I am now in command of the 580th ARC Wing."

Another low blow was handed the men. It was getting close to Christmas, and wing announced the Christmas leave policy, which said that anyone who had accrued leave could have a two-week leave over Christmas. Most of the men had used all their accrued leave for pre-embarkation. Anyone who didn't have any accrued leave could only have a three-day pass. By regulation, a three-day pass was limited to a 50-mile radius of the base, which meant that the men couldn't go home for Christmas.

I had a plan. I called in my first sergeant and told him I thought the situation called for drastic action—like breaking regulations. I told him my plan would only work if everyone followed my instructions to the letter. (I knew I was treading on quicksand.) I couldn't let all the men go at the same time, so I divided them into two groups and gave each man two concurrent three-day passes (an act against regulations). The first group would sign out on December 22, giving their home address. At the end of the first three-day pass, they would make a collect, person-to-person call to me at the orderly room phone number. The operator would say, "I have a collect call for Captain Saunders from Airman Smith in Chicago. Will you accept the call?" The first sergeant taking the call would say, "He's not here. Tell Airman Smith to call back tomorrow." The clerk would then sign Airman Smith in on the first three-day pass and out on the second pass.

The first group went out and came back, as planned. The next group went out, and all returned on schedule, except one man: a Staff Sergeant Coleman. I had warned all of the men that anyone not returning on time would be put on the Morning Report as being AWOL. This had to be done in case they had been in an accident. They were also warned that if they were put on the Morning Report as AWOL, they would be reduced one stripe.

When S/Sgt. Coleman returned five days late, I asked him his reason. He said he had heard that his aunt was sick, so he went to see her. I told him I was reducing him one stripe. He went to the legal officer to appeal my decision. The legal officer asked him to explain what had happened. He said he was on two three-day passes. When he said that, it "hit the fan." The legal officer was at wing, and as was the custom, the wing commander had to review all appeals. When he saw I had violated regulations by giving the men two three-day passes at the same time, he told the group commander to give me an administrative reprimand. The group commander, Lt. Colonel Williams, called me into his office. He said, "Sergeant Coleman went to Wing and appealed your reducing his rank by one stripe and said you gave all the men consecutive three-day passes. What in the hell were you doing?" I told him that I thought wing was insensitive to the needs of the men with the wing Christmas leave policy and that I thought I had done the right thing, considering the situation. He picked up a paper from his desk and said, "The wing commander told me to give you an administrative reprimand." He handed it to me, and I read it. He said, "I agree with you. Incidentally, that is the only copy. You can do whatever you want with it." Since it was wintertime (and cold) and the buildings were heated with coal-burning pot-bellied stoves—the colonel had one in his office—I walked over, opened the stove, dropped the reprimand in it, turned, saluted the colonel, and walked out.

The next morning at the roll call formation, I had the first sergeant call out the names of the men who were detailed to clean out the drain around the orderly room. I had told the first sergeant to put S/Sgt. Coleman on the detail. (I had been ordered to give S/Sgt. Coleman his stripe back.)At mid-morning, the first sergeant said that S/Sgt. Coleman wanted permission to go see the wing legal officer. I said, "Let him go." He went to the wing legal officer and told him I was harassing him and had him cleaning out a big ditch. The legal officer called and said he was coming to see me. He came to my office and said he had a complaint that I was harassing S/Sgt. Coleman and had him digging a big ditch. I took him outside and showed him the ditch the men were cleaning out. It was about 18 inches deep and about 2 feet wide. He shook his head, said, "Forget it," and left. I went to my office, called Colonel Williams, told him what had happened, and said that I wanted S/Sgt. Coleman transferred out of my squadron. He said he would take care of it. The next day, he was transferred to the motor pool.

In my squadron was a S/Sgt. Bird, who was the type of individual with a knack for getting into trouble. However, in some cases, it wasn't his fault. One evening he was drinking beer with some friends in Mountain Home, and two individuals from another table started trying to pick a fight with Sgt. Bird, who then left and went outside. The two men followed him, and a fight ensued. The military police came and cited S/Sgt. Bird for fighting in public. When the incident came to my attention, I checked into it and determined that S/Sgt. Bird was not at fault, so I took no action.

The next weekend, S/Sgt. Bird went to Mountain Home with some friends and returned to the barracks at about 10:30 p.m. The lights had been turned out in the barracks, so he turned them back on so he could see while getting ready for bed. According to squadron policy, the lights could stay on until 11:00 p.m. on Fridays and Saturdays. One of the airmen got up and turned the lights back off. S/Sgt. Bird turned the lights back on and told the airman to leave them on. When Bird went back to his bed, the airman turned them off again. When he did, S/Sgt. Bird went over and hit him in the face with his fist, knocking him down. The next day, the airman filed a complaint against S/Sgt. Bird for assault. S/Sgt. Bird was given a court-martial and sentenced to lose one stripe. I told him to appeal, adding that I would support him in the appeal.

S/Sgt. Bird appealed, and I was called to the wing commander's office. When he asked me about the appeal, I told him I thought the sentence should be reversed. He said that it looked as if Sgt. Bird was a troublemaker and that the report said he had gotten into trouble in town a few days before. I said, "I checked that out, and it wasn't his fault. He tried to avoid trouble with the other men. As far as the incident in the barracks, I thought because he ranked the other man, he gave the man a lawful order that should have been obeyed." The colonel said he agreed and told me to tell S/Sgt. Bird his stripe had been restored.

# Chapter XXVII

## A NEW JOB

They say all good things must come to an end. I guess all bad things must, too. Washington decided that when the wing deployed overseas, it would be a tenant on the base. Therefore, it would not need the support squadrons. Since my squadron was a support squadron, it was scheduled to be deactivated and the personnel transferred to other bases. The group commander told me I was being transferred to Carswell AFB in Fort Worth, Texas, the first time I had ever been assigned to a base in Texas. I was being assigned to the Flight Service Center, a tenant on the base. The base was a Strategic Air Command base, as well as the home of the B-36 Bombers. They had six pusher-type engines and were the biggest planes around at that time. It wasn't uncommon to see them come back from a training mission with one or more of the engines shut down and the props feathered (a procedure in which the prop blades were streamlined with the wind to stop them from turning).

The Flight Service Center controlled a large area, which encompassed all of Texas and parts of New Mexico, Oklahoma, Arkansas, and Louisiana. It had direct phone lines to all the bases in its area of control, as well as lines to adjoining Flight Service Centers. A base would file a plane's flight plan with us over their line, which also doubled as the Emergency Warning Net. We could open one or all the lines at one time to deliver an emergency message or request information on a specific flight. Periodically, we would be given notice to test the Emergency Net at a day and time. It could be daytime or nighttime. My job was as a control officer, and at the given time, I would open all the lines to the bases, call an Emergency Test, and tell all the bases to stand by and answer when their base name was called. We had to record the time each base responded and forward the report to Flight Service Headquarters.

While I was at Carswell, two significant things happened. One day I was on duty as control officer, and a call came in that a tornado had hit the downtown area of Waco, Texas. The caller said that there were many casualties and all types of blood were needed. I opened the lines to all of the bases and declared an emergency. I explained what had happened and asked all bases to check their areas for blood supplies and said I would call back. I got a very rapid response, and within thirty minutes, one of the bases had a C-47 airborne with blood for Waco. It proved the worth of the Emergency Net.

During my annual physical, I was told I would have to have a hernia operation. The next day, I reported to the hospital and underwent surgery to repair the hernia. The next afternoon, I was lying in my bed in the surgical ward, when suddenly the sky grew very dark and I heard a loud roar. It was a tornado hitting the base. I looked out the window and could see trees twisting down and the roof of the next building being torn off. A call came to the nurses' station for all medical personnel to report to flight line. Many men had been injured by flying debris, and the huge nose docks used to work on the B-36 engines had blown across the ramp and piled up in the corner of the fence, creating a pile of rubble.

As the nurse left, she turned to me and said, "You're in charge." I was the only officer in the ward. The wards were single-story wings off a common corridor. In the back of the ward were two doors. As I was looking at the destruction outside, a big gust of wind blew the double doors open, so I got up and managed to close them, but as I turned to go back to my bed, they blew open again. I threw up my hands to protect myself and caught the full force of the wind against the doors. It felt like I had torn all the stitches out where I had had the surgery. I felt sick at my stomach. I told two of the men to close the doors, and I went back to bed. When the nurse came back, she checked my incision and said it was okay.

Congress called for a Congressional Investigation to determine why the weather officer at the base hadn't known that the tornado was coming. How stupid could they get?

# Chapter XXVIII

## PARADISE

In 1953, I was transferred to Hickam AFB Hawaii, which was next door to the Pearl Harbor Naval Base. Hickam had a dock in the mouth of Pearl Harbor where tankers docked to offload fuel for the planes. It was a great place to fish for sharks, which would follow the ships, eating the garbage thrown overboard. When the ships came into Pearl, the sharks would follow.

To catch the sharks, we had a procedure that involved several steps. We would fish on the backside of the pier in shallow water, with small hooks and bread balls. There, we would catch small perch and fillet them, then change to a rod and reel, fish on the deep side of the pier, and catch a white eel, which we would then fillet and put on a larger hook and line. Afterwards, we would fish farther out and catch small Hammerhead sharks, which were about 2 feet long. We would switch to a deep-sea rod and reel with 40-pound test line and use the hammerhead for bait after cutting it with a razor to make it bleed. We would cast it out into the channel, where we would catch a 4- or 5-foot-long sand shark, which we would then use for bait.

We had two lines about 100 feet long: One line was a nylon rope, 1 inch in diameter, with a 10-foot chain tied to the end, with a U-bolt holding a large hook; the other line was a small-diameter steel cable, with a large hook clamped in the end. In the other ends of the lines, we had tied an inner tube from a C-124 tire. The tubes were hooked over the post used for tying up the tankers.

We would hook the large sharks through the tail, cut them to make them bleed, throw them out into the channel, and wait for action. We would catch a 10- or 12-foot shark. It would hit the end of the line, and the tube would pull it back. When the shark was tired out, we would go out in the 60-foot crash boat, shoot the shark, and tow it back to the boathouse. We would hook it onto the boatlift, lift it up, and put it on the pier. We would call a Honolulu merchant's number we had, and the merchants would come get the shark. It was quite an experience.

I was assigned to base operations as a clearance officer. One day about 4:00 a.m., a young captain who was the aide to General Maddoux, the Deputy Commander of Pacific Division, MATS, came in to file a clearance for Christmas Island. They were going there to fish. I checked the weather around Christmas Island, and it was all bad. Also, there were no radio aids there. I told the captain that, due to bad weather, I would not clear the flight. The young captain did not have clearing authority, so I had to clear the flight. He got very upset. He picked up the phone and said, "Do you want me to wake up General Maddoux and tell him you won't clear his flight?" He was trying to "pull the General's rank," an action I detested.

I grabbed the phone out of his hand and said, "Hell, no. Let me wake him up." He took the phone out of my hand, hung it up, and walked out. I didn't hear any more about it.

One afternoon, while I was working in operations, the crash alarm sounded. I went outside, and there was a C-124 (sans gear) on the runway. During bad weather, planes coming to Hickam would make an instrument let down on the Barbers Point Nav-aid and then proceed to the runway at Hickam. It seemed that the crew of the C-124 had been making practice letdowns without lowering the gear. When they decided to come in, they forgot to put the gear down. The crew consisted of a colonel from Division, the pilot, a lieutenant colonel, the copilot, a major, the acting engineer, and a lieutenant colonel navigator. The Honolulu newspaper said in its story that a contributing factor to the accident was too much rank in the cockpit.

I was transferred from operations to transient maintenance as the transient maintenance officer. I had civilian crews that serviced transient planes and performed minor maintenance on them. One night, a C-97 passenger plane with eighty passengers departed for the States. When it was about an hour out, the pilot called traffic control and said he was returning because he couldn't pressurize the cabin. He returned, parked at the terminal, checked with traffic control, and went home. The mandatory "crew rest" regulation wouldn't permit him to wait for repairs because he would exceed crew time. Operations called and asked me to see if I could get the plane back in service, as they had a backlog of passengers. I took a maintenance foreman in my jeep, and we went to see what was needed to put the plane back in service. When I drove up to the plane, I couldn't believe my eyes: the C-97 is a four-engine plane, and all the propeller tips on the two inboard engines had about 6 inches missing and all the tips of the two outboard engines were curled. The fuselage inboard of the engines had holes in it that you could throw a football through, where the prop tips had gone through the fuselage. I told the foreman to get into the jeep, and I drove out to the runway. I had an idea about what had happened. After takeoff, the pilot had settled back onto the runway. When we got to the runway, we found where the props had cut about 500 feet of the runway surface. I went to operations and told the traffic officer what I had found and to get the crew back out to operations. I called the tower and asked if they had seen anything unusual. When I told them what had happened, they said it had been sprinkling at the time and they saw what they had thought were raindrops reflecting in the landing lights. They said they guessed it could have been sparks. When the pilot and his crew returned to operations, I asked them to describe the takeoff to me. He said it was normal. I asked the engineer if his instruments, such as the torque indicators, had reflected normal operation. He said, "Yes." I asked the pilot if there had been any vibration during the takeoff or flight. He said there had been none. I told them to come with me, and I took them to the plane. I told the pilot that he had settled back onto the runway and cut about 500 feet of blacktop. He still insisted that he felt nothing wrong and no vibration. I told the engineer that the torque must have gone to hell when the props hit the runway. He claimed he saw nothing wrong. The pilot, copilot, and engineer were scheduled for proficiency evaluation flights.

They teach in aircraft maintenance schools that the props have to be perfectly-balanced, or vibration will ruin the engine. By all counts, the plane should have been a smoking pile of metal at the end of the runway. He should never have gotten into the air, much less flown an hour out and an hour back. It's like the landing at Reno. It's one of those things you just can't explain.

The air force pilots have always had the reputation of being womanizers. Back during World War II, the ferry pilots who ferried the planes from the factories to the bases in the States and overseas carried on that tradition. They were as bad as the navy crews, who had a woman in every port. They used to say that when a pilot kissed his wife good-bye at the front door, a ferry pilot would be coming in the back door. The Military Airlift Command at Hickam had flights to the bases in the Pacific, as well as flights to Travis AFB California. At that time, Travis was a Strategic Air Command, B-29 Base. SAC crews would be sent to bases overseas for training for as long as thirty to ninety days. Whenever the word went around that Travis crews were on overseas training, the MATS crews were fighting to get on the schedule to Travis. Some of the SAC wives were lonely, and the MATS pilots wanted to "comfort" them. There's a saying: "Absence makes the heart grow fonder...for someone else."

One of the MATS pilots, a friend of mine, told me this story. He had been making trips to Travis and "comforting" one of the SAC wives. On one trip, he got a room at the Officers Quarters, showered, changed into casual clothes, and phoned her. She said she would meet him at the Officers Club. They were having a drink, and he asked, "What do you hear from your Old Man?"

She said, "Oh, he's back."

My friend said, "Where is he?"

She said, "He's home babysitting."

My friend said he told her, "Let's get the hell out of here."

It came to the attention of the base commander that a captain and his wife were double-dating. They would go to the Officers Club, and he would look around for an unescorted woman, married or single, and take her home with him. Meanwhile, his wife would pick up one of the unattached officers and invite him home with her. The base commander had the captain transferred back to the States with a very low rating.

It wasn't just the air force personnel who conducted themselves this way. Some of the navy wives at Pearl Harbor (next to Hickam) were just as bad, especially the submarine crews' wives. The subs and some of the other ships would go to sea for extended periods of time, and some of the wives couldn't handle the loneliness. They were like the SAC wives.

One day a captain at Hickam reported that his thirteen-year-old daughter had not returned home the previous night, and he wanted to report her as missing. The provost officer reported the case to the Honolulu police's Missing Person Office. She was found two days later in a house in Honolulu, where she had been staying with a young man. The police turned her over to the provost marshal at the base. He called the girl's father and the pastor of the church her parents attended. They started questioning the girl, and she told this fantastic story. She said that in Hickam there was a club known as the "Non-Virgin Club," which was run by some young men and boys. The girls in the club ranged in age from twelve to fifteen. The boys would recruit young girls to be members of the club. During the war, the military built a number of reinforced concrete bunkers around the base, and the girl said that the boys would recruit potential members for the club, take them to one of the bunkers, and "indoctrinate" them into the club. When the base commander was told about it, he directed the base engineering officer to destroy all the bunkers on the base.

One day, I was informed that the next day I was to ferry a C-47 "Gooney Bird" back to Travis AFB. The next morning, I picked up the paperwork in operations. Unless the plane is empty, we always figure the load and balance. The manifest showed a lot of miscellaneous cargo. Also, a long-range fuel tank had been installed, as the flight to Travis was about 2,000 miles. According to my calculations, the plane was grossly overloaded. I closed out the weight and balance form at the maximum allowable weight and ignored the rest. I went into the weather office for a weather briefing. I remember the weather officer was a red-headed warrant officer who told me that I would have a little headwind on the first half of the trip and a good tailwind on the rest of the flight. He said I would be on top of all weather at 10,000 feet.

I visually checked the plane outside, kicked the tires, and said, "Let's go." I taxied out to the end of the runway, checked the engines, and got my clearance from the tower to take the runway. I lined up for takeoff and advanced the throttles, but instead of starting to roll, the plane just sat there and vibrated. It was so overloaded, it took it a while to start to roll. I finally got it in the air, and we headed for California. About halfway to California, there was a ship stationed called *Ocean Station November.* It is a navigational aid for planes flying between Hawaii and the mainland. When I got to the ship ahead of schedule, I should have known the weather briefing officer didn't know what he was doing. About halfway from the ship to the coast, we ran into bad weather and a strong headwind. On a flight like that, the pilot has to radio in a position report every hour. My navigator gave me our position, and I called it in to air traffic control. In a few minutes, they called back and asked me to confirm our position. I told my navigator that slip-horn control wanted us to recheck our position. He said, "That's where I said we were, and that's where we were." I called control and told them our position report was correct. The controller said the reason he questioned it was that, according to it, we had traveled only 100 miles in the previous hour. I told him we were aware of that because we were bucking a strong headwind and were loaded with ice. According to our weather briefing, we were supposed to be in the clear, with a good tailwind. I was beginning to be concerned about our remaining fuel. With the heavy load, we had burned more fuel than normal, and with the added weight of ice on the plane and the strong headwinds, it was

**114**

going to be close. Not long before that, the weather forecaster had busted his forecast for a flight to the States, and they ran out of gas and had to ditch the plane in the bay near San Francisco. We finally reached the coast and Travis AFB. When we parked on the ramp, I got out and went back to the tail of the plane. It was covered with ice, and when I hit the horizontal surface with my fist, cracks in the ice ran in every direction. Since our gas gauges had indicated zero, I had the engineer stick-check the tanks. He found just enough gas in one tank to wet the end of the stick. The other three wing tanks and the long-range tank were dry. That was too close. I hadn't used good judgment. I should have turned the plane down when I found it was overloaded.

I had to bum a ride back to Hickam. There was a navy plane going to Barber's Point, near Hickam. I asked the pilot if he had room for me. He said, "Sure. Come on." As we were going out to the plane, the pilot asked his engineer, "What did you do about that fuel leak?"

The engineer replied, "I stopped it up with paraffin. What do you think?"

The pilot said, "I think I'm tired of Travis. Let's go home."

So we went. What the heck? There was only 2,000 miles of water in front of us, and the paraffin would "probably" hold—and it did.

# Chapter XXIX

# RANK HAS ITS PRIVILEGES

General Sorry Smith was Commanding General of Pacific Division of MATS at that time. His wife was a "war bride" from England, who knew very little about the military. As wife of the commanding general, she was "Honorary President" of the Officers' Wives Club on Hickam. At one of the monthly meetings, a captain's wife came in late and apologized, saying she had had trouble getting her car started. Mrs. Smith asked, "My dear, why didn't you call a staff car?" She always got one when she called and, not knowing better, thought that a captain's wife could do the same, which showed just how little she knew about the military. She didn't know that a captain's wife couldn't get a staff car for any reason.

One day, I was at civil engineering, and one of the workers who came in was so scared that he was shaking. It seemed that Mrs. Smith had ordered civil engineering to make her an ironing board for her maid to use. They had gone to great lengths to make a very special board for her, one anyone would have been proud of. The man had delivered it, and when Mrs. Smith saw it, she chewed the man out unmercifully. It seemed that they had forgotten to put a cover on it. The civil engineering officer sent a staff car to Honolulu to get an ironing board cover for the ironing board. He didn't want to incur Mrs. Smith's wrath.

The general's pilot and aide was Major Johnny Gonge. When he wasn't flying the general on a fishing trip or an official trip, he was having the general's plane polished or taking care of Mrs. Smith's two poodles. A general's aide has to run errands or do whatever the general or his wife wants. If he is diplomatic and does a good job, he is rewarded with rapid promotions. One day, one of the poodles was sick. Instead of taking it to a vet, Johnny took it to the base hospital, where sick women and children had to wait while the doctor treated Mrs. Smith's dog. There's a saying in the military: "RHIP" [Rank Has Its Privileges]. When the general was transferred to Hamilton AFB California, Johnny went with him and continued to take care of the poodles. A few years later, I attended a "Hump Pilots" reunion at Harlingen, Texas. Low and behold, the guest speaker was Lt. General Johnny Gonge. I told him, "You have come a long way, Baby." I think if he had been trained as a vet by taking care of Mrs. Smith's poodles, he might have wound up a Five Star General.

During the Korean War, the wounded were flown back to the States. Hickam was a refueling stop. A butler hangar on the flight line had been converted to a waiting lounge for the passengers while they waited for the plane to be refueled. The Red Cross provided milk, cookies, and magazines for them while they waited. Red Cross volunteers from the Officers Club and Non-Commission Officers Club served the patients. Every year, when the Red Cross held a drive to raise funds for their needs, they always mentioned the hangar lounge, where *they* provided the milk, cookies, and magazines for the war wounded. They neglected to mention that the wives clubs were given a quota of cookies to bake, at their own expense, or that the milk was donated by the in-flight kitchen or that the magazines were donated by the passenger lounge at the terminal. The Red Cross never spent a dime on the hangar lounge; they just took credit for it. The "volunteers" even had to buy their uniforms from the Red Cross so they could serve. It's been my experience that this is typical of the Red Cross everywhere. In their drives, they also mention that they also provide money for the military men who have to go on emergency leave. What they fail to mention is that they require the men to sign a pay voucher for the money to be taken out of their pay and returned to the Red Cross before they will give them the money.

In West Texas, I know a case in which a small town was devastated by a flood that washed away a number of houses and cars and drowned several people. The Red Cross came and set up a shelter in the school, providing beds, food, and clothing for the homeless. Several of the affluent ranchers in the area gave the Red Cross checks of $100 or more as a means of thanking them for helping out. When the ordeal was over, the Red Cross sent the town a bill for "services rendered." The county commissioner's court held a meeting and asked all that had given the Red Cross money, or anything else, to report it to them. They had a photocopy made of the checks and made a list of other items that were given to the Red Cross, while they were providing the shelter. The county judge sent a letter to the Red Cross headquarters, with a copy of the checks and items provided, and asked them why. The Red Cross wrote back, saying that there had been a mistake: the city should not have been billed. I think that if the judge had not sent the letter, the "mistake" would never have been found.

While I was stationed at Hickam, I lived in a housing area known as Hickam Village. There were a lot of children in the housing area, and they had no place to play. I got permission from the base commander to use a vacant area in the housing area for a playground. I solicited help from some of the residents, and we went door to door, asking for funds to put up playground equipment and a fence. I set up a committee to assist in the planning and fundraising. To help raise more funds, we hired a carnival with games and rides on a percentage basis. We purchased the equipment, and some of the men helped install it and build a fence around the playground. When it was all completed, we had a grand opening, with ice cream and cold drinks for the children. I invited the base commander to cut the ribbon across the gate to officially open the playground.

There was a lieutenant colonel living in the village with his wife. They had no children (I think he hated children), and he was assigned to Seventh Air Force, which was over Hickam. He worked in the Personnel Directorate of Seventh Air Force and had been opposed to the playground project from the beginning. I found out later that the base commander had recommended me for the Commendation Medal for my work in providing the children with a safe place to play. Seventh Air Force was the approving authority for all awards. When the recommendation got to that headquarters, the lieutenant colonel had it disapproved because he didn't like my putting the playground near his house.

One thing that attracted tourists to Hawaii was the volcano on the island of Hawaii, or "The Big Island," as it was known. Before Hawaii became a state, people would say they were going to Hawaii when, in fact, they were going to the island of Oahu, where the city of Honolulu is located. The volcanoes erupted several times while I was stationed at Hickam. If it was a big eruption, I would get a copilot, and we would take a load of passengers to see the eruption. It was the most spectacular at night. The explosive eruptions would hurl molten lava high into the night sky, and there would be a river of bright red lava pouring down the side of the mountain and into the ocean. We had to be careful not to fly too low over the volcano or lava flows because of the thermal updrafts caused by the heat. The hot air rising would cause violent updrafts. When I was working in operations during an eruption, I would caution any pilot going to look at the volcano not to fly too low over the volcano or lava flows. One day a pilot took a load of passengers to see an eruption. The passengers were standing and looking out the windows when he flew low over the lava flow. There was a violent up-draft, which threw the passengers all over the cabin. The pilot's head hit the windshield, which cut a big gash in his forehead. Fortunately, no one was seri-ously injured or killed, and the plane didn't crash. A few days later, I saw one of the passengers I knew on Waikiki Beach. She had black and blue bruises all over her body and looked as if she had been beaten with a baseball bat.

While at Hickam, I was attached to the 50th Transport Squadron for flying. The squadron had C-124 aircraft, the largest transport aircraft in the air force at that time. I couldn't get over the fact that I was flying an aircraft that big. The check pilot who checked me out on the C-124 had a stack of 3"x5" cards on which were written the various problems pilots could encounter on the aircraft. When I was flying, he would hand a card back to the engineer. It might have said, "Wait five minutes, and say you have a fire in number three engine" or "Do nothing."

One day, when I was taking a check ride, the engineer said, "We have a fire in number three engine. This is for real." It was true. I calmly went through the procedure for controlling an engine fire—just as if it was a part of the check ride—and put the fire out. Later, the check pilot congratulated me on the way I had handled the emergency.

We got word one day that there had been a big earthquake in the Aleutian Islands and that a tsunami (tidal wave) was headed for the Hawaiian Islands. (A tidal wave will travel up to 200 mph when crossing the ocean.) I wanted to see what it looked like, so I got a C-47, and a friend of mine went as copilot. We went to see what it looked like when it came in. It was very strange. The shock wave on the bottom was in front of the shock wave on the surface. When the water wasn't very deep, we could see it churning up debris on the bottom. As it neared the land, the water seemed to run in all directions. Then the water would pull away from the shore and return as a huge tidal wave. When the water pulled back, it left ships sitting on the bottom of the harbor; when it returned, it washed boats a long distance inland. Tidal waves are quite destructive. The tidal wave was the last of Mother Nature's destructive forces I witnessed. I had flown through a typhoon on the way to Japan, been in an earthquake in India and several in California, flown through a tornado in Louisiana, witnessed the eruptions of a volcano, and experienced a tidal wave. I have found the saying "Nature in the raw is seldom mild" to be true.

# Chapter XXX

## THE AIR FORCE RESERVES:
## AN EMPTY EXTINGUISHER

I was at Hickam longer than at any other base, and it was the best assignment that I had had up until that time. However, all good things must come to and end—as I had received orders transferring me to Hamilton AFB California, located about 30 miles north of San Francisco in Marin County. The Fourth Air Force Headquarters was located there, but most of the flying was done by a reserve unit that was stationed there. My assignment was to check out the reserve pilots in the C-46 aircraft. Later, they changed to the C-119.

All of the reserve forces are considered a vital part of the combat capability of the military. They were supposed to fill designated positions in the active force in the event of a war, yet some of the pilots in the reserve unit at Hamilton were also pilots on a commercial airline. As such, they were expected to fly with the airline in the event of a war. The reserve unit was using them to fill up their manning slots, and the pilots were using the reserve to accrue points for retirement, with no thought of serving in the military in the event of a war. Several Congressmen also held positions in the reserve, and you know they wouldn't go into the military in the event of a call-up of the reserves. It made the reserve program very political. I found that no matter how bad a reservist was, it was almost impossible to put him out of the reserves. If someone tried to, they would write their Congressman, and the wrath of the Congressman would descend on that person, so it was best to accept it and "go with the flow."

The base had a C-47 that I could use when I wanted to go cross-country. One day I was returning to Hamilton from Salt Lake City, Utah. I had let down after crossing the Sierras and had just passed Sacramento when the cockpit filled with smoke. We couldn't determine the cause, so I called Travis AFB, which was just ahead of us, and asked for a straight-in approach. I told the tower that I had smoke in the cockpit. The tower said I should extend my base leg, as they had a plane practicing a "ground-controlled approach." I told the tower, "You get him the hell out of there. I have a cockpit fire, and I'm coming in to land." I landed, got it stopped, and got the six passengers out. I found that the soundproofing on the copilot's side of the cockpit was the source of the smoke. The copilot had been smoking, and the ashes from his cigarette had apparently set the soundproofing on fire. I took a portable extinguisher and put the fire out.

About ten minutes after we landed, the fire truck arrived. The chief arrived in a pickup. He jumped out and said, "Drop the battery." (The battery is under the belly of the plane, between the main gears. You open a cover plate and release a lock, and the battery comes down. Then you can disconnect it.)

I told him, "No problem. I put the fire out."

He said, "I said, 'Drop the battery.'"

I turned to my engineer and said, "Drop the damned battery."

He did, and the chief went inside, looked around, came out, got in his pickup, and left. If it had been a serious fire, by the time the fire truck got there, the plane would have burned up. I taxied into the line, went into operations, filed a clearance to Hamilton, and left—but not before I told the operations officer what had happened, how long it had taken the fire truck to get there, and what I thought of his tower operator and fire department.

The reserve unit changed planes from the C-46 to the C-119. I didn't have any time in the C-119 and couldn't be used as an instructor, so I was transferred to a Reserve Wing at Fort Miley, an abandoned coast guard facility on the peninsula not far from the Golden Gate Bridge. Some of the buildings were used as classrooms for the Reserve "Little Red Schoolhouse" Program in which the reserves would study various subjects at night on weekends. However, the program was worthless. In addition, they would serve two weeks on active duty each year, at an active air force base. For the most part, the classroom training had no connection to the job assignment they would have were they called to active duty. It was just a means of staying in the reserve program so they could retire when they reached retirement age.

Weed, California, is a little "one horse" town in Northern California near the Oregon border. It was in our area of jurisdiction, which included California, Nevada, and Utah. In Weed, there was a reserve unit consisting of five men who ranged in rank from airman third class to full colonel. They met on weekends and studied "basic management" and took time about being the instructor. When it was the airman's turn to be the teacher, he would be teaching a full colonel basic management, which was a waste of money and time. I recommended that the unit be deactivated but was informed that we were supposed to build up the reserve program, not tear it down. The situation was very political because the colonel had some friends in high places, and it took me over a year to get approval to close the unit.

My assignment at the wing was director of operations and training. All of the classroom training was under my control. In addition to the classroom training, there were two reserve groups in our area. There were two squadrons in each group, and they were located on civilian airfields where there were enough reservists in the area to man a squadron. The airfields were small civilian fields, which made the whole program rather stupid.

Whenever the mission of the reserve program was found to be unworkable, Reserve Headquarters in Washington would think up another mission, just to keep the reserve program going. The current program envisioned the reserve units recovering the aircraft returning from the first strike, in the event of a nuclear war. It was presumed that the aircraft, as well as the crew, would be contaminated with radioactivity. The reservists were to decontaminate the plane and crew and prepare it for a second strike. What made the mission so absurd was the location of the units on small civilian airfields. They could in no way accommodate a B-29 or B-52. The reserve units were just playing games.

To check on their "readiness," I would take a C-47 and fly to a field where a unit was meeting. I had obtained some low-level radioactive buttons from an active unit. They would be used to indicate radioactive contamination. We would hide some around in the plane and put one or two in our pockets to see if the ones with scanners could locate them. I would not tell the reserves that I was coming until we were about thirty minutes out. I would call them on the radio and give them a message, saying we were returning from a strike, had flown through an atomic cloud, was possibly radioactive, and had injured aboard. One or two of the crew would pretend to be injured when we landed. I would check to see if they found all of the buttons on the plane and on the crew and what procedures they used to decontaminate the plane and crew. A lot of it had to be simulated. I expected them to tell me what they would do if it had been an actual mission. Afterwards, we would hold a critique.

At Fort Miley, the coast guard had constructed heavily-reinforced concrete storage bunkers for the ammunition for the big guns and supplies. One day I told Colonel Parker, the wing commander, that the underground rooms would make a good bomb shelter and command post. They would be safe from anything, except a direct hit. With his approval, I wanted to take some of the reservists and make a project of it. He said, "Go ahead." I pointed out that the project would let the reservists utilize their military skills, instead of wasting their time in the classes. In the classes, there were reservists in just about every career field. It was no problem getting the ones I needed to help with the project. I could give them credit for their work on the project, the same type of credit they would have received by going to class.

Two steel doors opened into a corridor that ran the length of the bunker. The doors and the corridor were large enough for small trucks to enter, and rooms opened off the corridor. We obtained two large gas-operated generators and installed them in the end of the corridor, ventilated to the outside. In one small room, we walled up part of the door opening, plastered the walls, and made a water storage tank with an electric pump. In the event of a power failure, we could get water by gravity flow. In another small room, we installed a shower, lavatory, and commode. One room we filled with survival rations and canned drinking water. We made one room into an operating center with radio transmitters and receivers, as well as teletypes. In the last large room, we put twenty beds. This was a completely self-contained shelter. We had electricity connected to the outside. In the event that that was interrupted, we would use the generator. We had a ventilating system and estimated we could remain in the shelter for at least thirty days without venturing outside. The San Francisco paper sent two reporters out, and I took them on a tour of the facility. They took pictures and ran a good story about it.

# Chapter XXXI

## DIRTY POOL

Colonel Parker was an arrested cancer patient, and his wife had died of cancer. In the latter stages of the disease and at her death, she was cared for by the doctors at the base hospital at Hamilton AFB. Since we lived in Novota near the base, we used the base hospital for our medical needs. My young daughter Alana had been running a temperature for two or three days and had a sore throat. My wife called me at work and said Alana's temperature was over 103° F. I told her to take Alana to the hospital. When I got home that day, she told me that when she had taken Alana to the hospital, she was ignored for a long time. She finally got angry and demanded that the receptionist call a doctor to see Alana. When the doctor took my wife and Alana into the examining room, my wife proceeded to tell him what she thought about the treatment—or lack of treatment—for Alana. Her attitude made the doctor angry, and he told my wife to take Alana home and give her an aspirin. When I got home the next day, Alana was really sick. Her temperature was almost 105° F. I put her into the car and rushed her to the hospital. A different doctor checked her and said she had strep throat. I asked him how long he thought she had had it, and he said, "For several days." He entered her into the hospital.

One of the doctors at the hospital was a friend of mine. I went to him and told him what had happened. He said, "I'll tell you something if you promise not to tell where you heard it." I promised, and he said, "The doctor that saw your daughter the first time knew she had strep throat, but because he was angry with your wife, he didn't tell your wife or prescribe proper treatment." I felt like going to that doctor and beating some sense into his head, but I decided to handle it in the proper manner. I told Colonel Parker what had happened and that I was going to report the doctor to the inspector general for investigation of what the doctor had done. Colonel Parker looked up at me and said, "I wouldn't do that if I were you."

I asked, "Why not?"

He answered, "If you do that, I'll have to put on your OER (Officers Efficiency Report) that you can't get along with others."

I realized where he was coming from. Because the doctors at the hospital had cared for his wife before her death, he was threatening me with a very damaging OER if I caused the doctor any trouble. It didn't matter that my daughter had a serious illness and the doctor hadn't provided proper treatment for her because he was angry with my wife. He had me "over a barrel," and there was nothing I could do about it. An entry like that on my OER would ruin my career, and he knew it—and he knew that I knew it. I hated his guts for threatening me like that.

# Chapter XXXII

## ANOTHER COUNTRY, ANOTHER WAR

I had installed a teletype in the shelter and had it tuned to the news channel. I would take copies of the news to the colonel. One day after I had given him the news, he said, "You're in luck."

I asked, "How's that?"

He replied, "In the news you just brought, it said that the President of Vietnam has ordered all government officials to release their concubines."

I asked, "How does that make me lucky?"

He said, "Orders just came transferring you to Vietnam."

He handed me a copy, and it was true. I was being sent to Vietnam to Command Rescue Forces in Vietnam, Thailand, and Laos.

Soon I received a letter from a Colonel Walter Derck, Commander of the Pacific Air Rescue Center in Hawaii, which was over all rescue units in the Pacific. He suggested that I visit the rescue unit at Hamilton AFB to get an idea about how they worked and the way I would be operating. I was to find out later that the way they operated and the way I would be operating were as different as night and day. I would also find out that the center in Hawaii knew little about what was going on in Vietnam. In fact, the people in America did not know and, for the most part, would never know because of all the lies the Defense and State Departments were telling the American people about U.S. involvement in Vietnam. The American people and the world were told that we only had advisors in Vietnam who were advising the South Vietnamese government and military. In reality, the "advisors" were fighting alongside the South Vietnamese units they were "advising." What they were doing was advising the Vietcong with grenades and bullets that they shouldn't be in South Vietnam. Later, when we built up our forces in Vietnam, we admitted to fighting the Vietcong.

In the early stages of our involvement in the war, we had T-28 and A-1E single-engine, prop-driven aircraft over there, which were flown by American pilots and used to bomb and strafe enemy positions. To support the lie that they were only advising, a Vietnamese airman no class was put in the backseat, so if a plane was shot down, we could say that the Vietnamese was flying the plane and the American was only advising. Our helicopter gunships were also used to strafe and rocket enemy positions. They were only "advising." Another way we lied to the American people was through the news media. Every Friday, MACV (Military Assistance Command Vietnam) would hold a press briefing. Some of the troops called it the "Friday Follies." As an example of the subterfuge, the briefing officer would say, "In the past week, the following 'combat' losses were reported: 21 American, 200 Allies, and 800 enemy. Also lost to "combat" were one T-28 and two helicopters. They always reported high enemy losses. If you totaled up all of the enemy we reported we killed, we would have been guilty of genocide.

The losses were reported in such a way as to make the South Vietnamese military look good, which they weren't, and to not upset the American people with high casualty losses. What the press didn't know was that MACV had a rule for reporting losses. The trick phrase was "lost to combat." They had a rule that we had to find bullet holes in the body of a pilot for him to be considered "lost to combat." By the same token, bullet holes had to be found in a crashed plane for it to be listed "lost to combat." Most of the time, the only remains we would bring back from a crash site would be pieces of flesh. Since we were usually in enemy-occupied areas, we didn't stay around looking for holes in the parts of the plane. When we would return to the base, I would be asked if the plane had bullet holes in it and would respond, "I didn't see any."

They would say, "In that case, it probably had engine failure and was not lost to combat."

The same would hold true of the parts of bodies I would bring back. They would be listed as "not lost to combat." The reporters were not aware of the list of planes and personnel "not lost to combat." The subterfuge was used to keep down the losses reported to the American people. It was especially important when the college students started rioting against our involvement in the war in Vietnam. If they had known the real truth about what was going on in Vietnam and the real losses, they would have become even more violent. The American government and the Department of Defense have had a penchant for lying to the American people.

Another subterfuge the government and Department of Defense used was to give some losses a high-security classification to keep the information from the American people. They used the excuse that it was to keep the losses from the enemy. In most cases, the "enemy" was the American people.

# Chapter XXXIII

## CRIMINAL ACTS:
## THE RULE OF THE DELTA

One of the most criminal acts perpetrated against the Vietnamese people by the American pilots was their enforcing what the pilots called "The Rule of the Delta." A lot of people thought what Captain Calley did at My Lai—ordering the killing of women and children—was terrible. I think he was just following orders, and besides, what he did couldn't hold a candle to what the pilots were doing in the Delta. The Delta was the "rice bowl" of South Vietnam, and most of the year the fields were flooded for growing rice. Rice was the main staple in the farm families' diet, as well as the only source of income.

When the fields were flooded, the people traveled from their huts to the fields and to the nearest market by walking along the narrow dirt dikes that separated the fields.

The pilots, seeing the natives walking along the dikes, would let down to just above the dikes and fly straight at the people. They said the "Rule" was that if the natives jumped off the dike, they were enemy Vietcong, and the pilots would double-back and strafe them with their machine guns, killing many.

One day, when I was at Bien Hoa Air Base, I was talking to a group of T-28 pilots. A pilot who had just landed came over, and one of the group asked him what he had been doing. He said, "I've been enforcing The Rule of the Delta." The pilots all laughed.

I said, "I want to tell you something. You stupid bastards are killing friendlie's. Most of those poor people had never seen a plane till we came here. Naturally, they are going to jump off the dikes when they see a plane coming straight at them. I would like to get you down there and fly straight at you and see how long your guts would let your feet stay on the dike. If you jumped off, so help me, I would circle back and enforce what you call 'The Rule of the Delta.'"

### AGENT ORANGE

Another criminal act by the American Government perpetrated against many of the Vietnamese, as well as many Americans, was the spraying of Agent Orange over the countryside. Agent Orange was a strong herbicide that was used primarily to defoliate the jungle growth along roads and highways. This was done to prevent the Vietcong from waiting in the jungle growth along the roads and highways to ambush American vehicles.

However, Agent Orange was also used to kill the rice crops in the Delta. Some stupid individual up the chain of command had the misguided idea that the Vietcong were taking all the rice grown in the Delta to supplement their food supplies from the North. So, to prevent them from getting the rice, "Old Stupid" decided to kill all the rice with Agent Orange. It never entered his small brain that they would not only be depriving the people of their main source of food but would also be causing disease and death to the people and animals in the area sprayed. The long-range effect was even more horrible. Agent Orange caused terrible birth defects in both animals and children whose parents were subjected to the spray. I have seen children with terrible deformities caused by their mothers' having been sprayed with Agent Orange, and the United States government has done nothing to help them.

The spraying program was called "Ranch Hand," and the Government tried to keep it a secret that they were killing the rice. I saw a picture in the *Pacific Stars and Stripes* (I think) of three Ranch Hand planes spraying Agent Orange. The caption said that the planes were defoliating dense jungle to uncover enemy positions. The only problem with that was there were no trees in the picture: They were spraying rice fields! Whoever released the picture probably got into deep trouble because they didn't want word to get out that they were killing rice.

It has since been determined that Agent Orange has a devastating effect on animals and people. I think the Admiral responsible for sending Agent Orange to Vietnam got his pay for doing so: His son died from being subjected to the spray while serving in the Delta in Vietnam. The planes that were used to spray Agent Orange were parked on the ramp at the air base in Saigon. The trees and bushes were dead for about a half-mile downwind from the planes, just from the vapors of the spray on the planes. I worked in a building in the path of that vapor and was also subjected to the spray in the Delta. Since then, I have had sores break out on my neck, back, and shoulders. The ones on my neck bleed for a while and then go away for a month or two. I have asked the military dermatologists if the

*Looking for bodies at a crash site of a ranch-hand spray plane, shot down in the delta in Vietnam, 1963. Four bodies were recovered. Agent Orange was in the water of the field.*

sores were caused by Agent Orange, and they all parrot the same line: "Agent Orange doesn't cause any problems."

A major who was a doctor at the medical facility at Brooks Field was asked by a reporter about the effects of Agent Orange. He stated, "There are no known ill effects from being subjected to Agent Orange." I would like to take him and all the others who have said there were no ill effects from Agent Orange and put them in a field and spray them several times. They shouldn't mind. It would prove their point—if they were right. I think there was a conspiracy between the government and Defense Department to deny any problems with Agent Orange.

A medic at the VA Hospital told me I should be checked for Agent Orange. I made an appointment at the Audie Murphy V.A. Hospital in San Antonio, Texas. I ask the man who examined me if he was in Vietnam. He said he was. I ask him where he was stationed. He said he was stationed at Bien Hoa. I said, "I didn't know they had a hospital at Bien Hoa." He said he wasn't in a hospital. I ask him what he did. He said, "I was an ammunition loader." He was examining me for Agent Orange???

I filed a claim that I had been exposed to Agent Orange. I was sent a claim form to fill out. In the instructions, it stated if I was given compensation, a like amount would be deducted from my retirement pay. I threw the form in the trash. I still think all the "Nay" sayers should be sprayed with Agent Orange.

# Chapter XXXIV

## RESCUE OPERATIONS

When I arrived in Saigon, my predecessor in the rescue detachment had rotated back to the States, as had all the controllers. The stupid Director of Personnel at Rescue Headquarters didn't have the foresight to arrange for a short overlap of tours so that incoming personnel could be briefed on the situation and the method of operation, which was probably just as well in my case. My predecessor had been letting the combat operations section, which was in the next room, conduct the rescue mission, when a plane or chopper went down. In fact, shortly after I arrived, a chopper went down. I didn't know about it until after combat operations had sent other choppers to rescue the downed crew. When I found out about it, I went into the combat operations center and asked the American on duty why I had not been notified. He said, "Aw, go get a beer. That's what your predecessor did."

I found out later that a previous controller had been drunk on duty on more than one occasion. I told the captain there were going to be some changes. I said, "From now on, I want to be notified immediately when a plane or chopper is down, and I'll direct the rescue mission. Is that clear?" He said he understood.

The rescue center was in a small room in the combat operations building. It had one phone and no radios. There was no way I could conduct rescue operations with a setup like that. The first thing I did was to have the communications section install two telephones: one for the controller and one for me. I was assigned two captains as control officers and one sergeant to assist them. I later recommended that the sergeant position be changed to another captain. I then got another sergeant and a clerk. Next, I had radios installed in the center for contacting search aircraft or other stations. I had a metal shipping container placed alongside the building, and in it, we stored all the supplies we would need on a mission so we wouldn't have to wait for supplies in case a plane was down. We had helmets, grenades, parachutes, flare guns and flares, rifles and ammo, and my personal gun. I had brought a 375-caliber H&H Magnum rifle with me when I came to Vietnam. Before leaving the States, I went to Abercrombie & Fitch in San Francisco to get some ammunition for it. I told them I wanted 200 rounds of 300-grain solids. They asked if I was going to Africa for big game hunting. I told them that I was being sent to Vietnam and I wanted ammo that would shoot through the trees in the jungle. Twice, while I was in Vietnam, Abercrombie & Fitch had an officer who was coming to Vietnam bring me more ammo for the 375 at no charge. I carried an AR-15 when I went to the Delta and the 375 when I went in the jungle. In the Delta, I wanted the firepower the AR-15 would give me; in the jungle, I wanted something that would shoot through a tree and kill a man on the other side. The 300-grain solid in the 375 would do that. It was about equal to a 50-caliber round. The troops used a 50-caliber to explode landmines.

One day, the Vietnamese were sweeping an area so I could get to a crash site. They found five landmines but didn't have a 50-caliber gun.

We were in a graveyard. I told them to stack the mines on a grave mound and I would explode them. I moved off about 50 yards and got behind a grave mound. I aimed at the bottom mine, fired, and dropped down behind the mound. When the smoke and dust cleared, I could see a large crater where the mines had been and grave mounds had been blown flat over a large area.

An innovation that had been developed was a hand-held flare gun about 6 inches long and approximately the size of a fountain pen. The back end contained a spring-loaded firing pin, with a projection on the side that could be pulled back and released for firing or pulled back and locked in a safe position. The flare was in a tube about 3 inches long, with a cap end that screwed into the barrel. It could be held with one hand and fired. There were two or three companies that had come out with their version of the flare gun. I was asked by General Stillwell to test them and let him know which one I thought was best. The next time I went into the jungle on a mission, I took them and tried them out. With one of the guns, I could shoot a flare up through a hole in the jungle canopy that was no bigger than about 4 feet square. That was excellent, as some of the trees were as high as 200 feet. I told General Stillwell which gun I thought was best, and he ordered 800 from that company. When the company representative found out I had recommended his company's gun, he gave me a nickel-plated gun. I had also recommended that they make the flares in different colors so that a code could be made up. I used them so the choppers could follow me through the jungle.

Shortly after I took over the rescue unit at Tan Son Nhut Air Base, I was faced with a dilemma. I got a call that a parachutist's parachute had hung on the tail of the plane and he was being towed behind the plane. I went outside to see it, then went in and told the tower to have the plane circle the air base, adding that I would send a chopper to follow it in case the chute came loose. If the chute came open, the chopper could pick up the man. If it didn't open, the chopper could retrieve the body. I then had the pilot of the plane, who was Chinese, try different maneuvers to try to dislodge the chute, to no avail. The plane could not land without killing the parachutist. The only alternative that I could see was to try a daring rescue by flying up under the man, rising up to where I could catch his legs, pull him into the plane with me, and cut the lines. It would be very risky and could result in killing the man with the propeller of the rescue plane. He would die anyway, and it was a chance. I discussed my plan with the Director of Combat Operations and told him I wanted the best pilot of a T-28 plane that was available. He called me and said that a pilot from the field at Soc Trang would come to Tan Son Nhut for the mission. I went back outside, and the chopper was following the plane. I sharpened my survival knife so it would cut the cords fast. The mission would be risky not only for the parachutist but also for us in the rescue plane. I was sitting at my desk, going over in my mind how I would accomplish the feat, when the phone rang. It was the tower, and the man said that the chute had come loose and opened and that the chopper had picked up the man and was returning to base. Since then, I have often wondered if the plan would have worked. It would have been against great odds, but there was no alternative.

# Chapter XXXV

## NEVER READY FOR WAR

The U.S. has a bad habit of going to war before it is ready, training-wise or equipment-wise—and Vietnam was no exception. I was given the mission of providing rescue service, with no equipment. Whatever I needed, I had to scrounge.

When I took over the detachment, there were no radios in the control center—and no radios to take on a mission. When on the ground going to a crash, I had no means of contacting the Center or my aircraft overhead. Although the rescue detachment had been there for a year, Colonel Derck in the rescue center in Hawaii, Major Trexler, my predecessor, nor anyone else had done anything about the lack of equipment for the detachment. Later, I was able to get a single sideband radio for the detachment, as well as portable radios for field trips.

We also were not provided any helicopters. If I needed to go to a crash, I had to get an army chopper to take my team and me to the crash site. I also had no ground transportation. It was too far from the Detachment to the flight line to walk, so if I needed to go to the flight line to take a plane or chopper to a crash site, I had to hitchhike a ride, which would sometimes delay my departure.

I was able to scrounge a pair of jump boots. However, they were not suitable for wearing in the jungle or on hills. They had no tread or vents for letting water out. One day, I was briefing General Stillwell and his staff on rescue and jungle survival. During the briefing, he asked me what I recommended his chopper crews wear and what equipment they carry. He was wearing a pair of army jungle boots, which had tread on the soles, canvas tops, and holes along the sides to let water out. They also had steel insoles that prevented the spikes that the Vietcong had put in the trails from going through the soles and into the foot. He held up his foot and said, "What do you think of these?"

I answered, "I think they are great. I wish the air force had some. Sometimes in the jungle, when I'm wearing these jump boots, I take two steps and slide back one."

He then asked, "What size do you wear?" I told him, and he turned to a colonel and said, "Get him a pair." That afternoon, an army captain brought me the boots.

In the early stages of the war, the planes being used were old and worn out. The T-28 plane was a worn-out trainer. These planes were not designed to be used in a combat situation. There was a good reason to believe that some of the T-28 planes that had been lost were lost because the wings had come off during a dive-bombing run.

I went to the Bien Hoa Air Base to brief the pilots on jungle survival and rescue procedures. The pilots told me they were afraid to fly the old planes because every time they went up, they were afraid the wings were going to come off. They sounded like they were about to mutiny. I told them I would see what I could do about it.

Rescue Headquarters sent two helicopters with crews to a field on the Laos border in Thailand. The choppers were airlifted to the Udorn Air Base near there. I went there to check on the choppers and crews. The two pilots were captians, and one of them had been put in charge of the unit.

When I returned to Saigon, I went to see General "Buck" Anthis, the Commanding General of Second Air Division, which was over all air operations in Vietnam. I told him what the pilots had told me, and I thought they had a legitimate grievance. I suggested that the planes be given a thorough inspection before they were scheduled to fly again. I thought the pilots deserved that. He agreed with me and ordered all T-28 planes grounded until they had been inspected. During the inspections, several cracks were found in the spars, proving that the pilots' fears were well-founded.*

During the early part of the war, the troops had the old M-2 carbine rifle from. World War II. I recall my cousin—who was in the infantry in the Pacific during World War II—cursing the M-2, saying that when the men crawled through mud with the M-2 and fired it, the bolt would jam. Hitting the bolt with the butt of the hand to close it would cause the clip of ammo to fall out. Later in Vietnam, the old M-2 carbine rifle was replaced with the AR-15 rifle. Much later, the AR-15 was replaced with the M-16. I mentioned early that I got the first five AR-15 rifles that came to Vietnam for my team, and I had to wait for ammunition before we could use them.

They were using the old C-47 planes as flare ships to drop flares to disclose the enemy on the ground. I had flown the C-47 planes in India, twenty years earlier.

The military was never ready for war, equipment-wise. They always had "Vast Plans, with Half-Vast Support and Supplies."

*See the Mission of Captain Shanks

# Chapter XXXVI

# AREA OF RESPONSIBILITY

My area of responsibility for providing rescue of aircrews was the area of Vietnam below the DMZ, all of Laos and Thailand. In rescue operations, I was the "On Scene Commander" of all forces involved in the mission, whether they were air force, army, navy, marine, or Vietnamese.

My first priority was to familiarize my control officers and myself with the area of responsibility. I set up a field visit schedule for my controllers, as well as for myself, whereby we would visit every airfield and landing strip. We would collect data as to the size aircraft the fields could accommodate, facilities available, and security of the area. Although the detachment had been there for a year, my predecessors hadn't bothered to gather such data. We compiled the data into an Airfield Directory. It covered all of South Vietnam, from the DMZ in the North to Ca Mau in the Deep South.

On a trip to Ca Mau, I got a ride in a Carabou Army transport. The plane was to make stops at Vinh Long, Tra Vinh, and Ca Mau. I was looking forward to the flight—as I would be able to survey three fields. The pilot took off from Saigon and leveled off just above the treetops. I was quite concerned that we were flying so low. We were in range of small arms fire from the ground. When we landed at Vinh Long, I asked the pilot about flying so low. He said if he flew low, they couldn't hear the plane coming and we would be safer.

When we took off from Vinh Long, we did the same thing. The only problem with his logic was that we were flying across rice fields and tree-lined canals where the Vietcong could hide and see us coming. We landed at Tra Vinh without incident. Passengers sit side by side in the Carabou, and a captain was sitting next to me. When we took off for Ca Mau, I told the captain that if we got to Ca Mau without anything happening, we would win the "daily double." About 10 miles out of Ca Mau, there was a loud explosion. A rifle round had come through the fuselage between the captain and me and had exploded the nose wheel tire below the cockpit. Nevertheless, the pilot was able to make a safe landing with the flat tire.

I decided I had made my last flight with that pilot. I caught a flight back to Saigon with a different pilot. These survey trips and rescue missions put me on the ground in more of Vietnam than any other American—and probably any Vietnamese.

# Chapter XXXVII

## MISSIONS

At the outset, I formed a team consisting of two Explosion Ordinance Disposal (EOD) men and myself. I established a policy that I would lead my team to every crash site. I would augment the team with other specialists, such as mechanics, a photographer, etc. I wanted to keep the team small to facilitate getting into and out of a crash site in the minimum amount of time—to reduce the chances of an ambush. I always needed EOD men to destroy any bombs, ammo, guns, etc., to deprive the Vietcong of the ordinance.

In addition to the team, I would call on the Vietnamese Army to provide security forces. The size of this force would depend on the Intelligence estimate of the number of enemy in the area. There was always enemy in the crash area. The largest number of troops I used on a mission was two battalions; the smallest was twelve.

When a plane went down, I would make a reconnaissance flight over the area to select a landing zone as near to the crash site as possible, where the choppers could bring in my security force and my team. If there was enough daylight, I would try to complete the mission the same day the plane went down. It was important to try to get to the crash site before the enemy did. If there was

*Crash site of Captain Shanks, to the right of the "banana boat" helicopter.*

not enough daylight left, I would set up the mission for first light the next day. However, on one mission, the area was so enemy-controlled, it took three days to get the Vietnamese Rangers to agree to go into the area. On another mission, it took three days to secure the landing zone before I could take my team into the area.

## CAPTAIN SHANKS

Combat Operations notified me that a Captain Shanks had crashed in a rice field in the Delta. He had been dive-bombing a nearby Vietcong area, which for the most part was rice fields interspersed with canals. The Vietcong would build bamboo huts along the tree-lined canals, which made them difficult to see. Captain Shanks was bombing along a canal that was bordered with palm trees and undergrowth. The crash made a large crater in the middle of a rice paddy. All that was left of him were pieces of flesh that I dug out of the side of the crater. I put what I could find into my helmet liner and returned to the base. I turned the remains over to the morgue. I had established a policy that all bodies or parts of bodies were to be brought out and given to the morgue at Tan Son Nhut. The reason I brought back pieces of bodies was so the pathologist at the morgue could identify the individual, the mission could be closed, and there would be "No MIAs."

Captain Shanks had been writing his wife about the bad condition of the planes he was flying, telling her that he always worried that the wings might come off. I think that is what had happened on his last mission. When his wife was notified of his death, she gave his letters to *Life* magazine, and they were printed in a feature article. The magazine had a follow-up article in a subsequent issue, with pictures of his funeral in his hometown, with the flag-draped coffin.

What the military does in cases like this is get the individual's weight from his records and sandbag the coffin with that much weight. It is then sealed and not opened again. They can't just put sandbags in the coffin. There has to be part of the individual's body in the coffin also. Usually, when a plane impacts, the bodies of the crew disintegrate. Another reason I brought back parts of the bodies was that I felt a moral obligation not only to the pilots but also to their families to bring them back. As far as the pallbearers knew, his body was in the coffin, and as far as Mrs. Shanks knew, his body was home.

I had to submit a report of each mission to the Pacific Air Rescue Center (PARC) in Hawaii. After another mission in which I brought out parts of bodies, I got a letter from Colonel Derck, Commander of PARC (and my commander), criticizing me for bringing out parts of bodies. He said that when I got to a crash site and there was no one alive, I should leave and let someone else take care of the remains. Without exception, the crash sites I went to were in enemy-occupied areas, and we went into the area as a unit and left together, for safety reasons. No one left early. It was like the saying, "You hang together, to keep from hanging separately." Also, the only other individuals at the crash site were my security forces, and they couldn't have cared less about the remains of an American, unless it was one of my team or me. Besides, it was my policy to bring out all of the remains. I was told by a member of the rescue unit that after I had left the rescue unit in Vietnam and returned to the States, Colonel Derck made it his policy that if there was no radio contact with the pilot on the ground, rescue personnel would not even land. That could have resulted in not finding some of the MIAs we are looking for now but also in jeopardizing the life of a pilot who might have been too injured to use his radio—or the radio could have been broken when he hit the ground. I guess that Colonel Derck couldn't have figured that out.

One of my early missions was when a chopper went down in the Mekong River near the coast, and for two days, I had divers in the river looking for the chopper. We searched both sides of the river for about a mile downstream, looking for the bodies to no avail. The divers finally found the chopper quite a ways downstream from where it had impacted. The divers secured a cable to the chopper, and an H-21 dual rotor chopper lifted it out of the river. The bodies were still in the chopper.

## BA TO

The first major mission I conducted was the Mission to Ba To, a small native village about 30 miles from the city of Quang Ngai. It was in a mountainous area, inland from the coast. A B-26 had gone down in the mountains a few miles from Ba To. It was reported that a wing had come off on a dive-bombing mission, and the plane had crashed into the side of a mountain. The wing was quite a ways the side of the mountain.

When I had to go to a crash that was over 100 miles away, I used General Anthis's C-123 to take my team there. Quang Ngai had an airfield that I used as a staging area. It was also the headquarters of the Two Corps Commander. I took my team and flew to Quang Ngai. There was a U.S. Army chopper unit there. I contacted Colonel Byrne, the senior advisor to the Two Corps Commander, and requested an army chopper to take me on a reconnaissance of the crash area. The crash site was on the side of a mountain, and a canyon forked about a half-mile below the crash site. An area there could be cleared for a landing zone. I also asked Colonel Byrne to ask the Two Corps Commander to have a squadron of rangers proceed from the river, which was about 2 miles below the crash site, up the canyon to the area I had selected for a landing zone. They were to clear an area so we could land a chopper. We would proceed to the landing zone the next morning if we received word that it was clear.

The rangers found a Vietcong officers' housing area in the jungle in the canyon, about halfway to the landing zone. A firefight ensued, and the overwhelming force of the rangers drove the Vietcong up the canyon and over the mountain. There was a Vietcong village over the mountain. As usual, they carried their casualties with them. The Vietcong had some pigs, chickens, and buffalo there. The rangers killed everything, except the chickens, which they took to eat later.

That night, the rangers set up a bivouac in the canyon near the landing zone

*Crash site of B-26 near Ba To. Crash site (Arrow) and wing (Circle).*

area. They put out perimeter guards to prevent their being surprised by the Vietcong. During the night, one of the perimeter guards was killed and partially eaten by a tiger. The next day, I called in my report on the radio. I listed "1 EIA." I got a message back questioning the acronym "EIA." I sent a message back: "EIA, Eaten In Action." Everyone was familiar with "KIA" (Killed In Action), "WIA" (Wounded In Action), and "MIA" (Missing In Action); however, no one had ever heard of "EIA." This was a first.

The historian of the second air division wrote a story about the mission entitled "Mission To Ba To." I think the story may have been published in the *Pacific Stars and Stripes*.

The story follows:

## MISSION TO BA TO

HQ 2D DIV, TAN SON NUT AB, VIETNAM – At dusk one evening three haggard American airmen, in torn combat fatigues, walked out of the Vietnamese jungle near a place called Ba To. They carried with them a precious five-foot piece of metal. It was the wing spar of a B-26, which had crashed after the wing came off. The plane had crashed in dense foliage on a mountainside in Central Vietnam. The surrounding area was tangled brush and jungle where tigers roamed in search of food, and where the enemy Vietcong had one of his strongest concentrations. It was an area that no man entered unless he had to.

These men had to. The Air Force had to have that piece of wing. It had to know why the wing fell off. Until it did, the B-26, which was carrying the brunt of the load against the enemy Vietcong, was grounded. But to get that wing away from the Vietcong stronghold where it fell was as tough a job as any in the Vietnam War—and the type that rarely falls to airmen.

For 44 year old Major Alan Saunders, a rugged 6 foot Texan from Sonora, this was just another of the sort of mission which has earned him three Bronze Stars in Vietnam. Saunders was head of the small "Recovery" unit in Saigon when the call came in that a B-26 had crashed near Ba To. He immediately called for a C-123 to fly him and a four man team to Quang Ngai, about two-thirds the way up the East coast of Vietnam, and about 30 miles from the plane. The team consisted of Major John Kane, flying safety officer, Captain Carl Meek, SSGT Paul Merical, and SSGT James Harrington of the Explosive Ordinance Disposal unit. When they boarded the C-123 at Tan Son Nhut on 17 August 1963, Saunders and his men hadn't a clue what lay in store for them. In the next 48 hours, they were to make two trip's into the middle of one of the heaviest Vietcong concentrations in Vietnam and walk out again.

The V.C. were known to have been to the crash scene. In fact, it was later

*Briefing the chopper crew on the location of the landing zone near a crash site in Vietnam, 1963.*

learned that several V.C. had remained in the area observing the Air Force team while they went about their work the next day.

Saunders and his team were carried to the landing zone prepared by the ARVN [Army of Vietnam]. It was a small clearing in a grove of tall betel nut palm trees. The Army Huey pilot decided he couldn't take the whole team into the small palm fringed area with wind conditions as they were. He went back to the river, off loaded three of the team members, and made two trips taking the

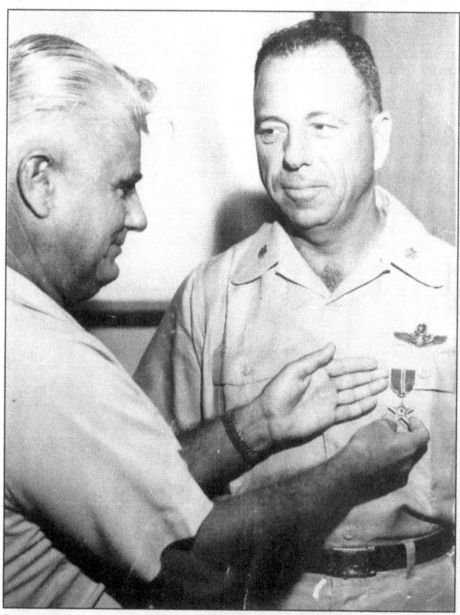

*Receiving my first Bronze Star for Heroism for my mission to Ba To. Colonel Allison Brooks made the presentation.*

team in. Each time he went in and out of the small clearing the UH-1B was subjected to small arms fire from the three ridges around the clearing.

Assembled on the ground, the team made its way to the plane. They passed bloodstains on the rocks along the trail, pointing the direction the V.C. took in carrying off their dead and wounded from the cantonment, which the ARVN had captured the day before. A village on the neighboring ridge seemed the logical destination.

The team reached the crash area around two o'clock, and picked up the bodies of the four men in the B-26, put them in body bags, and gave them to the ARVN to carry down the mountain. The Vietcong had already taken the parachutes and planned to get the rest later. Then they found seven 100 pound bombs, two 50 caliber machine guns and 150 rounds of ammunition, piled them on top the planes engine and wheel assembly and set a 40 pound charge of plastic explosive under the lot. They took cover behind a ridge near the plane. When the charge went off, it rocked the whole area. A forward Air Controller flying overhead saw Vietcong's running in all directions from the trees about 200 yards from the crash site.

On the way back to the landing zone, the team dodged sniper bullets but escaped casualties. The heavy sniping continued at the Army copter, which made three trips to the area, carrying out the team and the bodies.

Back at Quang Ngai, Saunders and his team drank some hot coffee and took

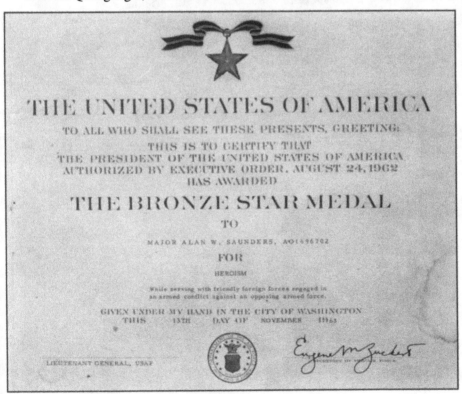

*The award of the Bronze Star medal for Heroism.*

a rest. A job well done, they thought. They were soon to find out that the real job had not yet started. A call came in to Saunders from 2nd Air Division in Saigon. "Go in and get the spar of the wing that broke off. We need it to find out the cause of the crash."

Saunders contacted the U.S. Army Senior Advisor in the area and again asked for a helicopter lift. The Army Colonel told him it was too risky and the chopper

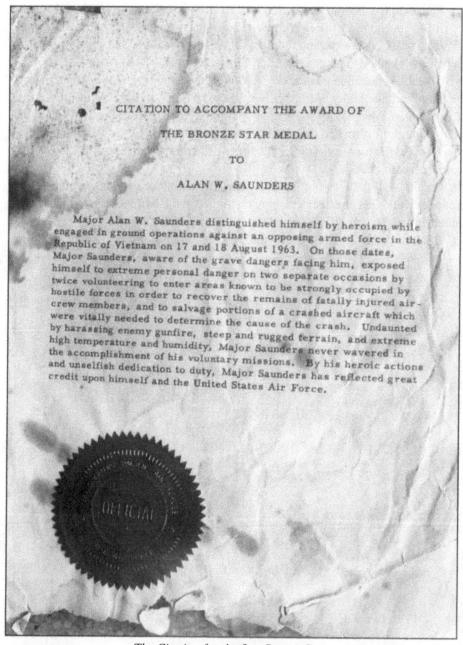

CITATION TO ACCOMPANY THE AWARD OF

THE BRONZE STAR MEDAL

TO

ALAN W. SAUNDERS

Major Alan W. Saunders distinguished himself by heroism while engaged in ground operations against an opposing armed force in the Republic of Vietnam on 17 and 18 August 1963. On those dates, Major Saunders, aware of the grave dangers facing him, exposed himself to extreme personal danger on two separate occasions by twice volunteering to enter areas known to be strongly occupied by hostile forces in order to recover the remains of fatally injured aircrew members, and to salvage portions of a crashed aircraft which were vitally needed to determine the cause of the crash. Undaunted by harassing enemy gunfire, steep and rugged terrain, and extreme high temperature and humidity, Major Saunders never wavered in the accomplishment of his voluntary missions. By his heroic actions and unselfish dedication to duty, Major Saunders has reflected great credit upon himself and the United States Air Force.

*The Citation for the first Bronze Star.*

142

would be lost if it tried to go into the same clearing again. The V.C. were in the area in strength, and probably expecting a return trip.

Saunders then asked if the chopper would take him just to the mouth of the Canyon on the river, 2 miles from the crash site. From there, his team would force march up the canyon to the plane's wing. This Colonel Byrne agreed to, but said he couldn't be responsible for the men's safety.

This time Saunders asked for volunteers, three men would join him in a highly dangerous mission. Every man on the team volunteered. Saunders picked Captain Earl Meek, SSGT Paul Merical and MSGT Hall, the latter being an expert on the B-26, who could determine where to cut the spar off.

This time, the mission had to be carefully planned so the team would have a reasonable chance of getting in and out of the crash area safely with the wing spar. First, additional plastic explosive and fuse had to be obtained. The plastic would be used as a shaped charge around the spar to cut it off. Some one-pound charges of TNT were all that was available. They were obtained from an ARVN Engineering Battalion that was building a road near Ba To. Getting good fuse to set the charge off was a different matter. Only old weathered fuse was available, and when several 8-inch lengths were tested, each burned a different length of time. Since a double charge had to be placed on the wing, and both charges had to go off simultaneously, the fuse had to be reliable. Saunders had made a call to Saigon to have some fuse flown to Quang Ngai as soon as possible. He received a call that the fuse had been located and would be flown to Ba To by noon. Saunders' plan called for departure of the team from Ba To at 14:40, 15 minutes to reach the landing zone on the river, an hour each way for the forced march, and an hour to do the job. To finish the job before dark, 14:40 was the latest the team could depart Ba To.

The plan for going into the canyon also called for a heavy aerial pre-strike of the area - with the exception of the immediate vicinity of the B-26 wing. Following the pre-strike, Vietnamese soldiers were to move in and provide a measure of security in the area for the recovery team.

Due to a crossing of signals, the ARVN moved in ahead of schedule and the aerial pre-strike had to be cancelled. This was a blow to the team. As airmen, they knew how much their safety depended on a heavy aerial pre-strike. The Vietcong would not hang around an area subjected to intense aerial attack.

Saunders decided to go ahead anyhow, as soon as the fuse arrived. At noon, it had not come. At 1300 hours, it still hadn't arrived. The H-21 helicopter that was to take the team to the mouth of the canyon, started up at 14:30. Still no bird coming in with the vital piece of dependable fuse. Loading the old weathered fuse and the explosives aboard the Army helicopter, the team left Ba To at 14:40 hours. Without the pre-strike, without the dependable fuse, and departing at the latest time possible, their job was going to be ten times harder but there was no turning back. With "Lady luck" on their side, and no delays, they would be able to accomplish their mission and return before dark.

At the mouth of the canyon, the team de-planed. They picked up as an interpreter, an ARVN Captain Loi, who had been trained at Fort Knox, and a security team of fifteen tough Vietnamese Rangers. The size of the security force had to be small to permit fast movement into and out of the crash site. The Captain told Saunders they would probably be ambushed going in, and would almost certainly be ambushed coming out. The Vietcong were on the ridge, and would be able to watch the team going in.

The group left the mouth of the canyon at 1455 hours, right on schedule, the 16 Vietnamese Rangers leading the way. The team crawled on their bellies across a brush-covered area of rocky ground, and here occurred another of those misfortunes that plagued the team from the beginning. Captain Meek from EOD fell over a rocky ledge and twisted his knee. Saunders had no choice but to send him back to the river, and 5 of the ARVN security force had to go with him. That left 11 to defend the team.

After an hour and 10 minutes of crawling through the rock patches and undergrowth that led to the plane's wing, the team reached their destination. Quickly, with Sergeant Hall directing, they cut away the aircraft skin, laid the charges, attached the fuse and then took cover in a washout down the mountain. For some unexplained reason, the Vietcong who were certainly watching the operation still hadn't opened fire.

The charge went off, but as Saunders had feared, one of the charges went off before the other and the job had to be done again. Risking their necks a second time, a new charge was laid, and this time, the precious piece of wing spar was separated.

The men rushed to the blast area, handed over the spar to the ARVN soldiers and then headed down the canyon. The Vietcong opened fire. Dodging sniper bullets and dog trotting in the open areas, the 3 airmen and 11 Vietnamese rangers made it back to the river in half an hour, 40 minutes quicker than the trip in.

Vietcong rifle and machine gun fire poured into the landing zone as the Army H-21 flew into the area to take the team out. With tracers flying all around the cockpit, miraculously the team was airlifted out without a single casualty.

Safely back at Ba To, the team rested before proceeding to Quang Ngai and then back to Saigon. The wing spar was sent to the rear depot for examination, and as a result of the findings, it was determined that the B-26's could once more take on bombs, effectively putting them back in operation.

For Saunders and his men, all of whom received the Bronze Star with V device for heroism, the episode was another day's work. Next day, he was back in his office in the Combat Operations Center at Tan Son Nhut, directing Recovery Operations. For his part, the Army pilot, Captain Woodmansee, was recommended for the Distinguished Flying Cross.

It's a strange war in Vietnam, and the men who fight it have to be prepared for anything. Saunders was trained as a pilot. He flew C-46's and C-47's over the "Hump" in World War II, and once spent two weeks behind Japanese lines in Burma trying to line up native support. This experience came in handy but for the mission to Ba To he had to learn a lot more. And the same with the men who accompanied him. They had learned what every airman assigned to Vietnam has to learn. Survival in a COIN war depends upon adaptability, and you have to write the book as you go along.

***

Before we left the landing zone on the river, I gave Captain Loi my AR-15 rifle. He was overjoyed. When I got to Quang Ngai, I asked Colonel Byrne to arrange for Captain Loi to have plenty of ammo for the rifle.

That night, I was drinking coffee with Colonel Byrne while waiting for the C-123 to arrive and take my team and me back to Saigon. Colonel Byrne paid me the greatest compliment I have ever received. He looked at me and said, "I want to tell you something. If I ever get into trouble over here, I hope you are there." A greater tribute could not have been made.

## LUCASSIG

Sometimes a pilot and plane are lost by following poor tactics. Captain Lucassig was a case in point. He was making low-level passes across open rice paddies at an enemy area along a palm-tree-lined canal. On the sixth pass, the Vietcong shot him down.

When the call came, I contacted the Army Chopper unit and set up two gun ships and a slick (a chopper without guns) to take me to the crash area. When I first went out on missions, I used two T-28 aircraft to escort my chopper. It didn't take long for me to realize that that was a mistake. They couldn't stay with the chopper, and while they were out turning around, I was getting shot at.

The pilot was flying in support of a two-battalion operation engaging a Vietcong force in the area. The ARVN had set up a command post in an abandoned house about a mile from the crash site. I planned to wait at the command post until the troops had secured the area around the crash site.

There was another native house about a quarter of a mile from the command post. As we approached the command post for landing, a Vietcong flag was run up on a pole outside the house, and the Vietcong fired on one of my gun ships. The door gunner returned fire. My pilot pulled out of the way and went in and landed at the command post, followed by the two gun ships. An American lieutenant colonel there was the advisor to the Vietnamese commander, and when we got out of the chopper, the advisor asked, "Who was doing the shooting?"

The door gunner said, "I was, Sir."

The advisor inquired, "What were you shooting at?"

The sergeant said, "I was shooting at that house. They ran up a Vietcong flag and were shooting at us."

The colonel said, "How many missions do you have, Sergeant?"

The sergeant said, "One hundred and twenty four, Sir."

The colonel took out a map and said, "Show me where it was."

When the sergeant did, the colonel turned to one of his captains and said, "Call Saigon, and have them send a couple of planes down here to bomb that house."

I said, "Colonel, you don't have to call Saigon. Let my gun ships take care of that house."

He said, "Okay." I told the pilots of my gun ships to destroy that house. After two passes with rockets, the house was gone—but the flag was still there on the pole.

That afternoon, the command post got a call from one of the battalion commanders indicating that they had secured the crash site and were standing by with four APC's (Armed Personnel Carriers). I told my team and the chopper pilots, "Let's go."

We were able to land in a rice field just outside of the tree-lined canal. The plane had cut a swath through the trees as it came down, shearing the wings off. The fuselage had come to rest on an embankment near a Vietcong house. The pilot's body was still in the fuselage. I looked in the house near the fuselage, and there was a large hole in the floor of the room. There was a woman and child in the hole. I helped them out of the hole and told them—in Vietnamese—to go. In looking around the area, I considered it easy to see why Lucassig was shot down. The Vietcong had holes in their houses for their families to get in and foxholes along the banks of the canal for the soldiers to get in when a plane was strafing or bombing the area. The only things Lucassig was killing were chickens, ducks, dogs, pigs, and cattle—anything above ground. The people were all in holes, and since he was coming in low level, his machine gun fire was not entering the holes. The Vietcong in front of him would stay down in the holes, and the ones on either side would stand up in the holes and fire at the plane. After about six

*Crash site of Captain Lucassig. Members of the team looking for his body.*

146

passes, they finally got him. We put his body in a body bag, gathered up the ordinance that was left, and stacked it on top of the wheel struts. The EOD members of my team put a charge of plastic under the pile and lit it, and we got down in the foxholes. After the explosion, we withdrew from the area and went back to Saigon. Upon arrival, I took Lucassig's body to the morgue and went to Division Combat Operations. I told the Director of Combat Operations what I had seen and suggested that when pilots bombed and strafed Vietcong positions along the canals, they approach at a steep angle so that the bullets would go into the foxholes. He ordered that change in tactics.

## GUTLESS VIETNAMESE PILOTS

I got a call that two army choppers had been shot down deep in the Delta, near the coast. As usual, I checked with Intelligence before going into a "hot" area. They told me that they had reports that the Vietcong had two 50-caliber guns in the area where the two choppers had been shot down. I called the army chopper unit for a chopper to take me to the crash area and asked Combat Operations for two fighters to escort us into the area.

We departed Saigon for the Delta and had gone quite a ways, when the pilot said he hadn't seen the escort. I told him to contact Peacock Control, and they would get the escort and the chopper together. The pilot called Peacock Control and was told that the Vietnamese Air Force had been given the escort mission, and the Vietnamese pilots had refused to go into the area because of the 50-caliber guns. My pilot asked me what I wanted to do. I told him to go to the crash area. I felt we had to go, with or without an escort. When we got there, a couple of army gun ships were still there, and they provided us cover, while we recovered the bodies from the choppers. The 50-caliber guns remained silent.

When we got back to Saigon, I delivered the remains to the morgue and returned to my office. I briefed my control officer on the mission and went into Combat Operations, which was a joint operation manned by both American and Vietnamese officers. I went to the VNAF (Vietnam Air Force) officer and asked him if he was aware that the Vietnamese pilots had refused to escort me to the crash site. He said he was aware of it. I asked him, "What are you going to do to them for not performing the assigned mission?"

He said, "Do? Do to them? We're not going to do anything to them."

I said, "Captain, I know you understand English, and I want you to listen closely to what I am going to say. If nothing is done to those pilots for not providing me escort and jeopardizing my safety and the mission, I will never again respond to a VNAF call for rescue. Do you understand me?" He just smiled. I don't think he believed I meant it. I turned and walked out, and I never responded to another VNAF crash.

The Vietnamese Air Force had a chopper rescue unit. However, it was known that their pilots would not go down to a crash site, or pick up wounded soldiers if there was any enemy ground fire. It was said that if the Vietnamese wounded saw that the chopper was a Vietnamese chopper, they gave up hope of being picked up. The VNAF pilots, like most of the Vietnamese soldiers, were gutless cowards. I think that's why they lost the war—that and the fact that the country was never put on a wartime footing. During the war, there were numbers of twenty- and thirty-year-old men driving taxis in Saigon. At noon, you didn't dare get in the street going off base at Tan Son Nhut, or you would get run over by the Vietnamese driving American jeeps, going home for a two-hour siesta. The Americans stayed on the job, running the war. Most of the Vietnamese officers had American jeeps, and I couldn't get a jeep or any kind of vehicle. I had to hitchhike, even to go on a mission.

## PILOT ERROR

Another unnecessary loss occurred when a U.S. Army plane carrying six people flew into the side of one of the highest mountains in South Vietnam. The plane was on a flight from Nha Trang to Buon Me Thuot. On a direct line between the two cities was the mountain. The normal flight path doglegged around the mountain. I couldn't understand why the pilot chose to go direct. Maybe he wasn't familiar with the area or just didn't look at his map. It was a fatal error. The plane impacted about 100 yards below the crest of the mountain, which was almost 9,000 feet high. Maybe he thought he was high enough to clear the top.

I took my team and flew to Buon Me Thuot. There we boarded a chopper that, with luck, would put us out on the top of the mountain. A company of ARVN troops was at the foot of the mountain. The side of the mountain, where the plane had impacted, had a steep slope down to the valley below; the other side had a sheer drop of about 4,000 feet, before sloping on down to the valley. We decided to approach the mountain from the cliff side. It was questionable if we could attain enough altitude to clear the top. Approaching the face of the cliff, we would know before we got to the mountaintop if we were going to clear it. There was brush and small trees on the summit, causing yet another problem. Would the chopper be able to hover at that altitude?

On the first attempt, we had to turn back. Not enough altitude. The second pass, we were able to clear the crest and hover. Two rangers rappelled down a rope and cleared an area for the chopper to sit down near the edge of the cliff. We offloaded our gear, and at the pilot's suggestion, we held the tail boom to steady it while the pilot started the chopper to keep it from tumbling off the cliff. We moved out from under the rotor wash, and the pilot lifted off, disappearing over the side of the cliff. The chopper came out about 3,000 feet below, looking like a small butterfly.

It was late by the time we reached the top of the mountain—too late to try to recover the bodies. I decided to camp overnight and recover the bodies the first thing the next morning. It was cold, windy, and wet on the mountain. We tried to start a fire to heat some water and food, but the wood was too wet. We were finally able to get a fire started by putting some plastic explosive under the wood. The plastic burns like gas and is not dangerous. We weren't expecting to remain overnight when we started out, and we didn't expect it to be cold on the mountain. I radioed out and asked that the chopper drop us some blankets. We had the six body bags, which are heavy plastic and zip up like sleeping bags. We could roll up in a blanket in the bag, zip it up, and stay warm.

Intelligence had advised me that there was a Vietcong company in that area and that we should check in hourly with the command post of the ARVN unit in the valley. They would need to know if we came under fire. I divided the night into two-hour watches and assigned a watch to each of the team members. I took the watch between 4:00 and 6:00 a.m. I thought that if the Vietcong hit us, it would be just before dawn. Each watch was responsible for waking up his relief. When the sergeant woke me, I unzipped the body bag and got up. It was so cold, I was shaking. When I checked in on the radio, my teeth were chattering, so they probably thought I was sending Morse code. I didn't think any place in Vietnam got that cold. The night passed without incident.

The next morning, we went down the boulder-strewn side of the mountain to the crash site. We recovered the remains of the six men, put them in body bags, and lugged them back up the mountain to the landing zone. I radioed the command post that we were ready for pickup. I told them we would need three choppers, or one chopper would have to make three trips.

When we got to Buon Me Thuot, the C-123 was waiting to take us to Saigon. Mission accomplished: no casualties.

## MISSING IN ACTION

Usually when a plane or chopper went down, someone saw it and reported the location. In this case, the pilot was on a mission alone and was reported missing when he failed to return to base.

The pilot of a B-26 was on a bombing mission southwest of Danang, reported over the target, and that was the last they had heard of him. I alerted my team and had the general's plane take us to Danang.

We arrived late, so I set up a search mission for first light the next morning. I was told how many planes and choppers I would have for the search missions, and I set up a briefing of the crews at 8:00 p.m. I took a map of the general area and marked it off in 1-mile squares. I numbered the squares outward from the last-known position of the pilot and briefed the search crews on their assigned search area the next morning. They were to strip-search from East to West across their assigned area, then North and South, paying close attention to the tops of the jungle canopy for evidence of burning. The problem was compounded because a lot of trees in that area were as high as 200 feet. A plane could crash through the trees and burn, not even scorching the tops of the trees.

We continued the search for seven days. Each area was searched by three or four different search planes or choppers. Every area was searched by a chopper crew, flying slowly at treetop level. The whole area around the target and a wide strip from the target to Danang were scrutinized to no avail. After seven days, I called off the search. The pilot was listed as "Missing In Action."

## DROP THE DAMNED LEAFLETS

As in the preceding mission, this pilot was flying alone. He was in a small single-engine aircraft. The pilot was a Forward Air Controller (FAC) and disappeared on a flight near Khe Sanh, which is in the northwestern part of South Vietnam, near the Laos border. I did not need my team, as there wouldn't be any ordinance to destroy if we found the plane. That area has dense jungle with high trees. It would be like looking for a needle in a haystack.

I flew to Danang and was briefed by the senior advisor, who said I would have three planes for the search. I decided to search for two days, and if we found nothing, I would cancel the search.

After two days of fruitless searching, I cancelled the search and flew back to Saigon. I was told that an army general in MAC-V wanted to see me. I went to his office, and he asked me about the search mission. I briefed him on the mission, and he asked me if we had dropped leaflets. I told him we hadn't, and he said we should have. I told him I didn't think it would do any good. He said, "I said drop the damned leaflets. I'll have USIS [U.S. Information Service] print the leaflets." The next day, I went to USIS to get the leaflets. They had printed one million, which were about 2"x4" in size. On the front was a likeness of a FAC plane; on the back was [in Vietnamese] "If you have seen a single engine plane—like the one on the other side of this leaflet—go down in your area, you should notify the local police." It was so stupid, it was comical. It reminded me of the saying, "Forgive them, Lord, for they know not what they do."

In the first place, the natives in that area are very primitive: They do not even have a written language. They are born, live, and die in the jungle. It was doubtful they had ever seen a plane of any kind. As for reporting it to the police, there are no police in such a primitive society.

I took the leaflets to Danang, flew to the search area, and dropped the "damned leaflets" as directed. It showed how little the general and the USIA (who printed the leaflets) knew about the situation in Vietnam. The general was not alone in that regard. There were many like him in high positions in Vietnam who didn't know what the hell they were doing.

## VC VALLEY

There is an area west of Danang (near the Laos border) known as "VC [named for the concentration of Vietcong in that area] Valley." The Vietcong owned the place. A river runs through the area, with high hills on both sides. Jungle growth borders both sides of the river and covers the foothills. Some of the trees in the jungle are 100- to 150-feet high.

Two T-28 fighter/bombers were on a bombing mission near the river when one of them was shot down. The pilot of the other plane saw the plane crash and then disintegrate. He did not see a parachute, which would indicate that the pilot had bailed out successfully.

When the pilot of OK-Grad-1 reported that his wingman had crashed, I was notified at my office at Tan Son Nhut. I called the Air Division Operations to have the general's crew stand by to take my team and me to Danang. I called EOD and asked that Sergeants Merical and Harrington report to the flight line with explosives and equipment for a mission in I-Corps. I asked Major John Kane of the Second Air Division Flying Safety Office to go along also.

With the team assembled, we boarded the C-123 and headed for Danang, arriving after dark. Colonel Benjamin Preston, Senior Air Force Advisor to I-Corps, met the plane when we landed. He said, "You have more problems than you had when you left Saigon."

I said, "What's happened now?"

He said, "The marine squadron commander wouldn't wait for you. He sent two marine choppers and a FAC to the river in VC valley, with twelve men on board. They landed on a gravel bar on the river just at dark."

The FAC pilot said, "I saw their lights as they lifted off, and they disappeared, probably shot down." The FAC had given Colonel Preston the coordinates where the choppers had disappeared.

By authority of the Joint Chiefs of Staff, the commander of the rescue unit is to be the "On Scene Commander" of all forces involved in a rescue mission. The marine commander should have waited for me. It would have saved the lives of the twelve marines lost, as well as the men lost trying to recover their bodies.

With Colonel Preston's help, I got two companies of Vietnamese rangers as a security force. I set up a heli-born operation for first light the next day to take the rangers to the area where the marine choppers had disappeared to secure the area for a search for the missing. The Vietcong knew we would come looking for the choppers, and they would be concentrated in the area.

At daybreak, we launched the operation. As the first chopper came into the landing zone, enemy fire wounded the copilot and killed a ranger seated behind him. As more choppers came in, two more marines were wounded. The FAC pilot who was controlling the landing had a fuel line shot into and his Vietnamese observer wounded. The pilot was able to glide out of the area and make a forced landing in a safe area. The troops were pinned down all day. That night, about midnight, I was in radio contact with the troop commander. He said they had only secured about 200 meters of ground, had dug in, and were taking fire from all four quadrants. The next two days were spent flying out wounded and dead rangers. I had T-28's, B-26's, and gun ships bombing, strafing, and rocketing the Vietcong positions. They finally withdrew, and the rangers were able to get to the crash site of the choppers. They could only find ten of the twelve men who had been on board. In my daily report, I listed the two as "Missing In Action."

During the time from when I got to Danang till the bodies were removed from the crash site, the marine squadron commander was very offensive and obnoxious. He was the living example that there are more "horses' asses" than there are horses. I think it was because he had screwed up by sending the choppers out that evening, and the marine commander, a colonel, probably had chewed him out for doing it. The sad part of it was that where they had set down on the gravel bar was about 5 kilometers downstream from the T-28 crash site.

I radioed the troop commander and told him to send half of his troops up the river to secure an area around the crash site. The rest of the troops were to provide security around the landing zone. I had two marine choppers take my team and me to the landing zone on the gravel bar. I felt sure the Vietcong had booby-trapped the trail up the river, so I decided to cut a new trail. It took us several hours to reach the crash site. We found a wing and some parts of the plane in the edge of the river. The rest was either in the river or on the other side. We could not cross the river because it was flooding due to heavy rains in the area. We searched the bank for any signs of the pilot—to no avail.

There was a clearing near there, but the Vietcong had erected poles around the area to prevent choppers from landing. I told the troops to take the poles down. I radioed the marine lieutenant colonel and asked him to send two choppers to the landing zone we had cleared of poles, to take the team back to Danang. The bastard said that the choppers were not going to come up there because it was too dangerous. Furthermore, if we were not at the gravel bar by five o'clock, he was going to leave me out there. At that point, he was more of a threat to my team and me than the Vietcong were. If he had been in range, I believe I would have shot him.

We had to cut short the time for searching for the pilot and started back to the landing zone on the gravel bar. If we were left in Vietcong Valley overnight, I didn't think we would survive. An ambush on the return trip was probable and expected. We retraced our route back to the landing zone. Major Kane twisted his ankle jumping a ditch and needed help walking, which was going to slow us down. I radioed the marine lieutenant colonel and told him of our problem and that we would be delayed. He said, "I'll give you another thirty minutes. If you are not at the LZ by 5:30, you will be left there." We traveled as fast as we could, with two men helping Major Kane. We arrived at the landing zone with about five minutes to spare and could hear the first chopper coming in.

I told the troop commander to keep the troops in the edge of the jungle, and I told Major Kane that he and Sergeants Merical and Harrington were to go out on the first chopper. Sergeant Merical and I helped Major Kane out to the gravel bar, and they boarded the chopper. I moved out from under the rotor wash, and the chopper lifted off. It got about 20 feet in the air, and the Vietcong shot it down. The men on the chopper ran for the jungle. I took cover in a hole in the gravel surrounded with rocks that had been used as a guard position at night. We started taking fire from across the river and from the hills. I started returning fire with my 375 Magnum rifle. The troops were returning fire with small arms and grenade launchers. I had two T-28's and four gun ships overhead, and they were also receiving fire. They called me from the edge of the jungle and said that Major Kane and the copilot were missing. I thought they were probably dead in the chopper. I got up and ran to the chopper, but there was no one inside. I called Major Kane, and he answered from some brush nearby. I ran to them, and the copilot and I helped Major Kane into the jungle. I ran back to the chopper to see if I could tell which way the fire came from that shot it down. It looked as if the

152

shots came from some jungle growth up the side of the mountain on our side of the river. The T-28s and choppers were bombing and strafing the Vietcong who were firing at them. I ran back to the jungle and told my radioman to let me have the mike. I told the FAC overhead to tell the choppers to expend on the jungle growth on the side of the mountain. I watched as they started circling the area. The lead chopper started firing tracers, and when the tracers started hitting the jungle growth, he fired rockets. Each chopper, in turn, repeated that procedure. After about twenty minutes, a period that seemed like hours, the Vietcong stopped firing. Another marine chopper came in. The pilot said he was to only pick up the crew from the downed chopper. However, when he landed, there was a rush for the chopper, and I think eleven men left on it. Two more marine choppers came in and took my team out. An army gun ship came in and picked me up. Before heading for Danang, we made one more pass with rockets and machine gunfire on the jungle growth on the side of the mountain. For good measure, I fired another magazine of bullets from my 375 into the jungle growth. When I was pinned down on the gravel bar, I fired it so many times and so fast that the bolt was burning my hand.

I felt that the marine chopper getting shot down was just retribution for the marine lieutenant colonel refusing to send the choppers to the landing zone up the river because, as he said, "It was too dangerous."

*Award of the third Bronze Star for Heroism for the second mission to VC Valley.*

**153**

It was Sunday, and the General's C-123 crew was waiting to take me back to Saigon. Every Monday morning at nine o'clock, General Anthis held a staff meeting at Second Air Division Headquarters. The department heads took turns briefing the staff on their operations the past week and any problems they were having. When it came my turn, I briefed the staff on my mission to VC Valley

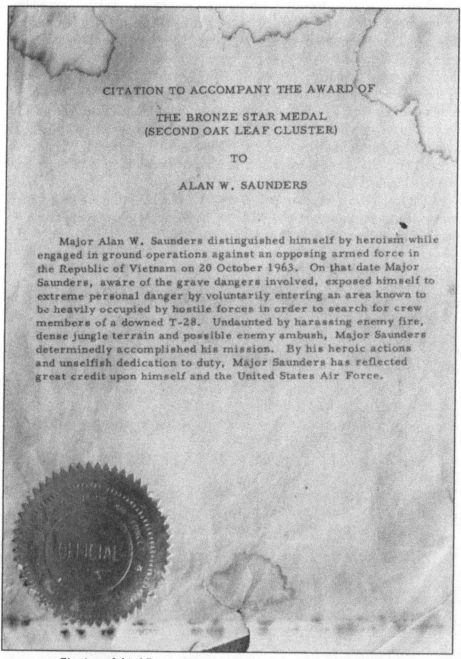

CITATION TO ACCOMPANY THE AWARD OF

THE BRONZE STAR MEDAL
(SECOND OAK LEAF CLUSTER)

TO

ALAN W. SAUNDERS

Major Alan W. Saunders distinguished himself by heroism while engaged in ground operations against an opposing armed force in the Republic of Vietnam on 20 October 1963. On that date Major Saunders, aware of the grave dangers involved, exposed himself to extreme personal danger by voluntarily entering an area known to be heavily occupied by hostile forces in order to search for crew members of a downed T-28. Undaunted by harassing enemy fire, dense jungle terrain and possible enemy ambush, Major Saunders determinedly accomplished his mission. By his heroic actions and unselfish dedication to duty, Major Saunders has reflected great credit upon himself and the United States Air Force.

*Citation of third Bronze Star for the second mission to VC Valley.*

154

and how rotten the marine lieutenant colonel had treated me. I recommended that the mission be closed and the three men, two marines, and the pilot on Grad-2 be listed as Missing in Action. General Anthis complimented me on what I had done and said, "I want to see you in my office after the meeting."

After the meeting, I went to his office and wasn't prepared for what he said: "You have to go back up to the crash site."

I asked, "Why? There have been too many losses already."

He said, "You didn't find any part of the plane connected with the pilot." I didn't relish going back to that hellhole. "I'll tell you about an incident that happened to me during the Second War," he said. "I was in Africa, doing the same kind of work you are doing. A C-47 was missing in the jungle. We conducted an extensive search for several days and found no trace of the plane or people on board." I told the general I thought we should cancel the search. He said, "Fine. You go pack your bags."

I asked, "Why?"

He said, "You are going to the States to tell the relatives of those missing how hard you looked for their loved ones, or you can go look some more." I got the message.

I decided not to take my team with me. I alerted the crew of the General's C-123 to take me back to Danang. When I arrived there, I went to see Colonel Preston and told him that General Anthis wanted me to go back to VC Valley and search the opposite side of the river for the pilot of Grad-2. He said, "I would like to go with you."

I said, "Great. I would like to have you along." I told him that the river was at flood stage and we would need a lot of rope and a rubber raft, like the ones we had carried on over water flights. He said he could arrange that. I also told him that I didn't want to have anything to do with the marines. I wanted the army choppers to take us in and bring us out. He agreed. I had briefed him on Sunday about how the marine lieutenant colonel had treated my team and me.

I set up the flight to the upper landing zone for an eight o'clock departure. The troop commander of my security troops had told me just before I left the landing zone on Sunday that they were going over the mountain the next day to hit a Vietcong outpost. I radioed him to take a company of his troops to the crash site and secure the area by 8:00 a.m. the next day. The next morning, Colonel Preston and I, along with the rope and raft, were taken to the upper landing zone by an army chopper.

The rope was stretched across the river. The rangers pulled themselves across on the rope. Colonel Preston and I were taken across in the raft. The troops started to take up positions near the river, when a firefight started with the Vietcong. The troops were able to force the Vietcong to retreat, giving us a chance to search the area. The Vietcong continued to snipe at us as we searched. I found the small plastic air vent from the cockpit canopy and put it in my fatigue pocket so I could show General Anthis that I had found a part of the plane that was close to the pilot. We searched the area and found no trace of the missing pilot. As some of the plane was in the river, it was possible that his body had been washed downstream but even more likely that his body had disintegrated upon impact, and because several days had elapsed before we got there, animals had consumed the pieces. The troops got into another firefight, so we withdrew back across the river. I radioed the pilot of the army chopper and told him we were ready to be picked up. Colonel Preston and I were lifted out of the landing zone without further incident and returned to Danang.

I called my office in Saigon and gave Captain O'Connell the information for the daily report and told him to send the C-123 for me the next morning. That night, Colonel Preston gave me a pencil sketch he had made of the area around the crash site and had named it "Saunders' Place."

The next morning, when I arrived back at Tan Son Nhut Air Base, I briefed General Anthis on the result of my second trip to VC Valley. I showed him the air vent I had recovered at the crash site. He agreed that I should close the mission. It was the only mission that I had conducted with losses. The losses were needless and not my fault. They were the result of the marines' stupidity. There is a fine line between bravery and stupidity, and in this case, the marines had crossed that line.

## HISTORICAL MISSION

About ten years after I retired from the Air Force, I received a call from the Air Force Office of Information. The caller said that they were writing the history of the Air War in Vietnam. They wanted to include a rescue mission and had selected one of mine. When he described the one they wanted to use, I recognized it as the mission to VC Valley. They wanted my permission to write about me and wanted any input I could give them to supplement the information they had gleaned from the air force files. I sent them some facts about the mission, and they wrote the story.

The air force's story "When We'd Only Just Begun" follows:

## "WHEN WE'D ONLY JUST BEGUN..."
## AN EARLY RESCUE ATTEMPT, 1963

"Beatles" were bugs, "pot" was something you cooked in, and "Jolly Green Giant" referred to a can of corn. In 1963, if we thought of Vietnam at all, it was only to confirm our commitment to stop "them" over there before we had to fight "them" in California. It was a simpler, happier time of idealistic vision—a time when we'd only just begun.

Many young Americans saw their visions become illusions in the mud, blood, and dust of Southeast Asia. For them, the symbolic torch of freedom, which an idealistic young president had taken up in the name of a new generation, was finally extinguished in the rotor wash of the last helicopter evacuating Saigon. But those same helicopters, the huge Sikorsky HH-3E Jolly Green Giants and the larger HH-53B Super Jolly Green Giants of the Aerospace Rescue and Recovery Service will always be remembered with great admiration by those of us who served in Southeast Asia. We assumed that these gargantuan contraptions, flown by their dedicated aircrews, had always been available to "bear any burden, meet any hardship, support any friend, oppose any foe..." so that others may live.

During World War II, it became the policy of the United States Army Air Forces and, later, that of the U.S. Air Force, that those men who fly and fight would never be abandoned—not to suffering, not to death, and not to capture. The Air Rescue Service, founded on these principles in 1946, received its bap-

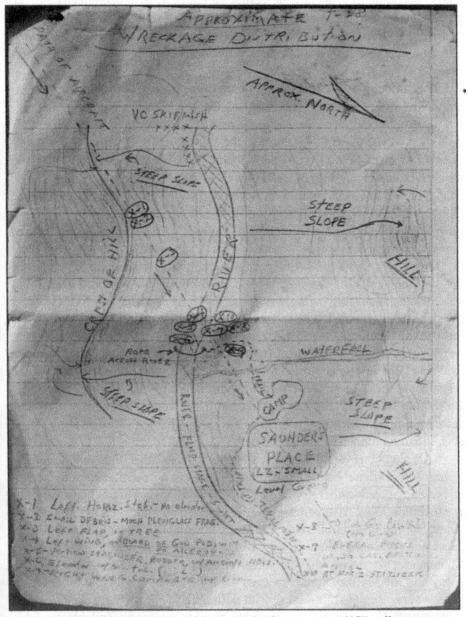

*Colonel Preston's map of the upper landing zone in " VC" valley.*

tism by fire during the Korean War and emerged from that conflict with a legendary reputation earned in the performance of daring aircrew recoveries far behind enemy lines.

When the smoke cleared from the battlegrounds of Korea, the United States found its new responsibilities as the Free World's strongest defender. American planes, stationed around the globe, stood ready to strike at the two foremost communist powers: the Soviet Union and Red China. The Air Rescue Service conformed to this commitment by adopting the "Global SAR [Search and Rescue] concept, under which the peacetime roll of SAR was theoretically extendable to wartime requirements." Helicopters were of limited use under an aegis that required Rescue vehicles capable of flying missions compatible with long-range bombers. By early 1960, there were virtually no helicopters in the ARS inventory.

The Indochina War spanned a generation, finding its roots in the anti-colonialist struggle of the Vietnamese against the French, even before the area was overrun by the Japanese early in World War II. President John F. Kennedy directly involved the U.S. Air Force in the conflict in 1961, when he sent a new breed of Airmen, the Air Commandos, to the Republic of Vietnam (RVN) to help train the expanding Vietnamese Air Force (VNAF). Under the code names "Jungle Jim" and "Farm Gate," these men risked their lives in aging, propeller-driven Douglas B-26 "Invader" bombers and slow-moving North American T-28 trainers specifically modified for a combat role. Indeed, many training situations placed these men in a shooting war.

In response to the increasing pace of air activity, Detachment 3, Pacific Air Rescue Center (PARC) was established at Tan Son Nhut Air Base on April 1, 1962. Detachment 3 manning was set at three officers and two NCO's. Their mission was to establish a rescue control network for coordinating SAR efforts throughout Southeast Asia. Helicopters manned by ARS personnel, however, were needed to rescue downed aircrew members, and these helicopters were not available. The rescue controllers at Tan Son Nhut were forced to rely on the resources at hand: Farm Gate fixed wing aircraft, a few U.S. Army Piasecki H-21 "Flying Bananas," and Marine Sikorsky H-34 helicopters. The VNAF possessed a limited helicopter inventory that was available for rescue work; however, it had no trained rescue personnel.

SAR situations varied greatly with geography, weather, time of day, and the disposition of friendly and enemy forces. However, a general pattern for search and rescue operations developed early in the Vietnam conflict. When an aircraft was overdue at its destination, this was reported to Det 3, PARC at Tan Son Nhut. The rescue controllers then requested available aircraft to search for the crash site. As soon as the wreckage of the missing plane was found, the Det 3 Commander marshaled whatever forces he could muster for the SAR effort. The Army and the Marines agreed to make their helicopters available for SAR, as long as such use did not interfere with normal missions. Usually, American resources were secured for rescue attempts, but if they were unavailable, the SAR

Commander had to rely on Vietnamese helicopters. In most cases, the Commander of Det 3 flew to the immediate area by helicopter if it was located within 100 miles of Saigon. If the aircraft was down farther away, he took whatever transportation was available, typically a C-47 or C-123, to a base near the site. A rescue force was then assembled and transported by helicopters to a clearing near the wreckage. Occasionally, a ground force from the Army of Vietnam (ARVN) was required to secure the area before the SAR force could be set down. After reaching the downed aircraft, the survivors or the bodies were carried back to the helicopter-landing zone (LZ).

Some of the problems encountered during rescue attempts in the period before the advent of the HH-3E's and the HH-53B's were illustrated in a rescue operation that occurred in October of 1963. On the afternoon of October 8, under the call signs of Grad OK 01 and Grad OK 02, a pair of Farm Gate T-28s rolled in on a target in the mountainous jungle area near the Laos border west of Danang. The American pilot of Grad OK 02 bore in low on his bomb run, unaware that it was to be his last. The pilot of Grad OK 01 watched as his wingman's aircraft went out of control near the termination of the dive and disintegrated upon impact. The pilot of Grad OK 01 returned to base and, at the mission debriefing, reported the aircraft down in the jungle near a swollen stream that flowed into the Buong River. Major Alan Saunders, Commander of Detachment 3, was notified at his office in Saigon.

Alan Saunders, a tough, six foot tall, 44-year-old Texan from Sonora who regularly carried a 357 Magnum* [sic] on his forays into the jungle regarded this as just another mission—not unlike one that had already earned him a Bronze Star Medal in Vietnam. Saunders gathered his rescue team, a 2-man explosive ordinance detail (EOD), a flying safety officer (FSO), and a photographer. Within an hour, they were on a C-123 bound for Danang.

When Saunders reached Danang, he assumed the role of on-scene commander (OSC). He was told that two U.S. Marine H-34's had proceeded to the crash site to search for survivors shortly before dark and had disappeared in an area known as "V.C. Valley." This upped the ante.

At dawn, Marine helicopters lifted two companies of ARVN infantrymen to the area of the downed aircraft. As the helicopters were landing, enemy troops firing from the surrounding hillsides wounded three U.S.Marine crewmen and killed one ARVN soldier. Farm Gate T-28's B-26's and VNAF propeller A-1 attack planes responded by strafing the enemy positions. A USAF L-19 light observation plane directing the strike aircraft took a hit that punctured a fuel line and wounded the VNAF observer. The American pilot nursed his crippled aircraft with its suffering Vietnamese observer back to a forced landing in friendly territory. Meanwhile, the ARVN force landed and began hacking out a larger LZ to facilitate future landings. When the task was finished, the troops started working their way to the site of the H-34 crashes. They reached the downed helicop-

---

*375 Magnum rifle*

**159**

ters the next morning, October 10, only to find the remains of ten of the twelve persons who had been on board the two aircraft. The other two, if they survived, were probably carried off by the Vietcong.

On 11 October Major Saunders secured Marine helicopters to airlift his party to the landing zone. Extremely heavy fire from a force situated on a ridge line near the LZ, forced the pilots to divert to a gravel bar that extended into the stream about five kilometers south of the T-28 wreckage. Once on the ground, Saunders led the party through the dense jungle underbrush to minimize the risk of encountering booby traps often planted along jungle trails by the Vietcong. The party hacked its way through the jungle for several hours before reaching the partially immersed wreckage of the T-28. They found a half-submerged wing at the edge of the swollen stream and a horizontal stabilizer on the opposite bank. The remainder of the debris was either under water or strewn out along the opposite shore. While the photographer snapped pictures, the FSO, Major John Kane, cut a spar from the wing for later structural analysis to determine the cause of the crash.

The Marine helicopter crew, not wanting to fly into VC Valley after dark, radioed Saunders, instructing him to be back at the gravel bar no later than 1700 or risk spending the night in Vietcong territory. Still fearful of enemy booby traps or ambushes along the jungle trails, the weary party started cutting a path through the tangled underbrush, back toward the pickup point. Major Kane sprained his angle jumping a ravine and had to be carried, further slowing their progress. With the hot sun sinking toward the horizon faster than the rescue party was moving toward the gravel bar, Saunders was forced to radio a request for a thirty-minute extension of the rendezvous deadline. Reluctantly, the Marines agreed, and the exhausted men struggled on through the stifling heat and binding bush, reaching the gravel bar with only minutes to spare. The first helicopter appeared overhead at exactly 1730. Major Kane and the EOD members climbed aboard along with several ARVN soldiers. The H-34 had risen only a few feet when Vietcong automatic weapons fire ripped into the engine. The chopper settled back onto the gravel and the passengers and crew scurried to cover in the underbrush along the bank. While six Army Bell "Huey" helicopters raked the ridge line with their machine guns and a pair of Farm Gate T-28's pounded enemy positions with bombs, Major Saunders crept back to the disabled H-34 to check the damage, and determined that it could not be flown. As soon as the hostile fire was silenced, two H-34's darted in and snatched the members of Saunders' beleaguered team to safety. Just before dark, an Army "Huey" landed and picked up Saunders. The bulk of the 120-man ARVN company spent another night in VC Valley before being airlifted out at first light.

Meanwhile, Saunders and his team returned to Saigon to check on business at Det 3 and file reports on the SAR efforts in connection with Grad OK 02. On 14 October, he flew to Danang once again, this time to organize a search party to return to the T-28 crash site and attempt to find a clue to the fate of the crew. This time Saunders arranged for a Green Beret team to lead an ARVN Ranger company, which he had obtained to provide escort. At dawn, a Marine H-34 carried Saunders, the Green Berets and the ARVN Rangers back to the LZ slashed out of the jungle in connection with this rescue effort. Light Vietcong resistance was encountered, slowing their progress so that the T-28 wreckage was not reached until the following day. The unexamined portion of the debris rested on the opposite side of the swollen stream, so the Green Berets led the Vietnamese Rangers across the swiftly flowing water. Vietcong held their fire until the party emerged from the stream and then let loose with full fury. Under the guidance of the Green Berets, the Vietnamese moved forward and soon dispersed the enemy troops as they secured the riverbank. Nevertheless, the persistent Vietcong continued sniping at the rescuers until dark, but the night passed quietly. Then, as the sun broke above the distant mountains, the guerrillas mounted a savage attack that forced the ARVN to take up defensive positions on a ridgeline overlooking the crash site. Heavy fighting continued for several hours before the area was finally secured and Saunders was able to organize an intensive search for the missing pilots. After three days, no sign of the crewmen was found. Saunders at last decided that either the Vietcong had disposed of the two men, or they had been carried away by the fast moving waters of the flooded stream. A helicopter then extracted Saunders and returned him to Danang.

Major Saunders flew back to Saigon, a bustling, burgeoning city filled with intrigue, vice, and a deepening American involvement. Around the bars of the Continental Palace and Caravelle, where American advisors shared their scotch with French journalists, Hong Kong businessmen, and the Intelligence agents of several countries, Alan Saunders was known as "Jungle King." For Jungle King and his men, this episode in the Annamese wilderness was "all in a day's work." These harrowing experiences were not unusual for the brave and dedicated Air Force man who served Air Rescue Service on the front lines of freedom. In those early days of the growing American presence in South Vietnam, Jolly Green Giant referred to a can of corn, and A-1 "Sandy" escort and "Crown" C-130 control ships were unknown entities in an unimaginable future. But, the rescue spirit— the driving force that urged men to risk their lives so that others may live—permeated every challenging day—even back then, when we'd only just begun.

## FOOTNOTES

1. John L. Vandergrif, Jr., ed. *A History of the Air Rescue Service*. (Winter Park, Fla., 1959), pp. 75-82.

2. Vandergrif, *A History of the Air Rescue Service*, p. 94; and Ltr, USAF to MATS, Subj.
   Reorganization of ARS, dtd, 26 Sept. 1958.

3. *A History of the Air Rescue Service (MATS)*, 1 Jan.-31 Dec. 1960, p. 13.

4. Office of Air Force History, *The United States Air Force In Southeast Asia, Illustrated*, 1976, Government Printing Office, pp. 12-18.

5. Memo, *Farm Gate Activity Report*, 13 Feb. 1962, pp. 2-3.

6. Msg (C), Det 3, PARC to Hdq ARS, 181535Z Oct. 1963.

Msg (C), Det 3, PARC to Hdq ARS, 201710Z, Oct. 1963.

Msg (C), Det 3, PARC to Hdq ARS, 231010Z Oct. 1963.

## MISSION TO "C" ZONE - COLONEL HERGERT

The area of III-Corps was divided into lettered zones. One thing that every-one agreed on was that "C" Zone belonged to the Vietcong. There hadn't been a "Friendly" in that area for years. That was to change.

On a Sunday afternoon, I was resting at home when the gate bell rang. It was the Air Police to tell me that I was needed at my office. A plane was missing. I told him to wait and take me to the base. When I got to the rescue center, the duty officer said that a T-28, flown by a Colonel Hergert, Deputy Chief of MAGAF (Military Advisory Group Air Force), was down in C Zone.

I don't know why a man of his rank and position was flying a strike mission over C Zone on a Sunday afternoon. I think he was either trying to get a medal or qualifying for "combat pay." They had changed the rules on qualification for combat pay. You had to be "subjected" to enemy fire to qualify. Furthermore, in his position as DC/MAGAF, he knew too much to risk possible capture by the Vietcong.

The FAC that was with him when he went down provided the coordinates of the crash site. I arranged for a gun ship to take me to the area to select a landing zone near the crash site. The plane had crashed near a small stream. The whole area was covered with dense bamboo. The nearest clear area for a landing zone was about a half-mile away. It was late when we returned to the base.

*Ready to depart in a gun ship to look for a crash site and L.Z.*

I decided to set up a heliborn operation for first light the next morning. I wanted a pre-strike of the landing zone before the choppers got there. I called the American Senior Advisor to the ARVN and told him I would need security troops the next morning for a mission to C Zone. I thought two companies of Airborne Brigade would be enough. He said he would contact the ARVN commander and get back to me. Later that evening, he called and said the ARVN troop commander had said he wouldn't go into C Zone with less than a regiment and none was available. I passed that information on to the army command post and advised them that we would not be able to go to the crash site at first light, as I had planned. All the next day, negotiations continued with the ARVN troop commander. The American general, who was Head of MAGAF, was getting very upset that we had not been able to go to the crash site. That night, the ARVN troop commander agreed to give me two companies of Airborne Brigade. I laid on the pre-strike and troop movement for first light the next morning.

It took a fleet of helicopters to move the troops and me the next morning. I had always taken Sergeants Merical and Harrington from EOD with me on all the missions, except the last one into VC Valley. When I called EOD for them the evening before the mission, I was told that the director of material, who was over EOD, said they could not go to C Zone. "It was too dangerous."

It was just getting light in the East when we boarded the choppers for the landing zone. I went in one of the H-21's (Banana Boats) with some of the troops and the troop commander who could speak English. The H-21's were escorted by six gun ships. When we got to the landing zone, the pre-strike had not taken place. It had been given to the VNAF, and as usual, they had screwed up. We had to pull out of the area and wait for the VNAF T-28's to strike. Our coming into the area and then pulling out told the Vietcong that we were coming.

The VNAF T-28's finally got there and struck the area around the landing zone. Following the pre-strike, we landed. When I had selected the landing zone, I had noted the compass direction from the landing zone to the crash site and told the troop commander the way to go. The area was covered with large, dense bamboo, which had small, dense bamboo underneath. This made movement very slow and dangerous. I could only see about 10 feet in front of me. An ambush was almost a certainty. I would move about 10 feet, stop, and listen. I had arranged with one of the gun ships to follow me overhead and stay in radio contact in case we were ambushed—and to keep me on the right path to the crash site. It placed him in more danger to enemy fire than we were. He completely disregarded his safety and followed my movement to the crash site and back to the landing zone.

On the way to the crash site, I had troops ahead of me, behind me, and on both sides of me. As I was moving toward the crash site, I suddenly started meeting the ARVN troops instead of following them. A captain came along with his radio operator. I said, "Captain, where the hell do you think you are going?"

He said, "We're going back to guard the LZ." I didn't have the time or the patience to put up with any cowardly Vietnamese troops.

*Dense bamboo between the landing zone and the crash site of Colonel Hergert.*

I reached over and got a handful of the front of his fatigues and said, "Captain, I know you understand English. You get on that radio and turn these troops around. You are going to the crash site." He talked to his troops on the radio, and they started going forward again toward the crash site. It puzzled me that the Vietcong had not ambushed us.

Because of the dense bamboo, it took about an hour to cover the distance from the landing zone to the crash site. The Vietcong had been there and had piled up some pieces of the plane down by the stream. The plane had impacted almost vertically. The engine was in the hole it had made on impact. One of the wings was nearby. I checked the wing and was surprised to see that there were holes from shell fragments that had penetrated from the top of the wing. That meant that Colonel Hergert had been shot down by an airburst. Up to that time, Intelligence had no knowledge that the Vietcong had weapons with that capability.

Colonel Hergert's body had disintegrated on impact, and I searched the area for parts of his body. I found part of his left hand, with his wedding ring still on his finger. I put the parts in a "burp bag" (air sick bag), and we made our way back to the landing zone.

I decided to have a chopper take me into Tay Ninh. I would wait there until we got word that all the troops were safely out of the landing zone before we left for Saigon.

I had always been concerned that someday the pilot and the copilot of a chopper I was in would get hit, either in the air or on the ground. I wanted to be able to get home. While we were waiting to go to Saigon, I asked one of the chopper pilots to show me how to start and fly a chopper. I told him the reason. He said, "Okay." He showed me how to start it and let me take it up, fly around, and come back in for a landing, which was the hard part because I had trouble getting it stopped. However, I felt I knew enough to get me back in an emergency.

While we were waiting for word that the troops were all out, a radio call was received that there was a chopper coming from Saigon, and we were to wait for its arrival before leaving for Saigon. The chopper arrived and started unloading boxes. I wondered what was going on and asked the pilot what was in the boxes. He said that Secretary of Defense McNamara was in Saigon and had been told that I had gone to C Zone to get Colonel Hergert's remains. He said to send us some good food. There was fried chicken, fried potatoes, and even ice cream. I couldn't believe it. This coming from the Secretary of Defense.

Word came that the troops were on their way back to Saigon. I took Colonel Hergert's remains, and we headed for Saigon. When we parked on the ramp, a staff car came up, and a captain got out. He said to me, "I've come for Colonel Hergert's remains."

I said, "I'm sorry, Captain. I don't give remains to anyone except the mortician at the morgue."

He said, "I work for the General. Colonel Hergert's wife is on the base, and I want his remains."

I said, "Captain, if you wanted them so damned bad, why in the hell didn't you go get them?"

He got back in the staff car and left. I took what was left of Colonel Hergert to the morgue. I often wondered how the captain would have reacted had I handed him the burp bag containing what was left of Colonel Hergert.

Secretary of Defense McNamara, having been briefed on what I had been faced with in going into C Zone to get Colonel Hergert, paid me the greatest tribute by sending the ice cream and fried chicken. The recognition and tribute by the Secretary of Defense meant more and was a greater tribute than any medal.

The army chopper crew, flying just above the tops of the bamboo, followed me into the crash site and back out. They disregarded their own safety to ensure mine. For that, I recommended them for the Distinguished Flying Cross, which they were awarded.

# Chapter XXXVIII

## FIRST OUT-OF-COUNTRY RAID:
## A COMEDY OF ERRORS

The U.S. Navy had a carrier task force on station, off the coast of Danang, South Vietnam. The task force consisted of one of the large carriers and its supporting ships. The planes from the carrier were flying reconnaissance missions over the "Plain of Jars" in Laos.

One day, I received a call that one of the planes had been shot down over Laos. The carrier had sent planes to look for the pilot. The next day, one of the search planes was shot down by the Pathet Lao forces. The Pathet Lao forces were supporting the Vietcong who were going through Laos on the Ho Chi Minh trail on their way to infiltrate South Vietnam. This compounded the problem. Now, we had two planes and pilots to look for.

Washington was following the reports of the operation. That evening, Second Air Division received a message that said Washington wanted a list of three potential targets on the Plain of Jars in Laos. They would select one of the targets for an air strike. They wanted to show "them" *what we could do* if they shot down our planes. Second Air Division was told to set up a strike mission and wait for a target.

This turned out to be the most ill-planned, ill-briefed, ill-executed bombing missions of the war. Mistake number one was when Washington decided to let eight F-100 Fighter Bombers from Clark Air Force Base in the Philippines fly the mission. They should have let the navy carrier have the mission. In the first place, their pilots knew the area. They had been flying reconnaissance missions over the area. Also, it was their planes that were shot down, and they wanted retribution. Mistake number two was that they let Colonel George Lavin, Commander of the Jet Unit at Clark, lead the mission. I have never figured out why—in the Air Force—when there is a special mission, some colonel or general wants to lead it. Maybe, like Colonel Hergert, Colonel Lavin wanted a Medal or combat pay. Mistake number three: The mission should not have been controlled from Washington. They were too far removed from the scene. Keep in mind that this was the first out-of-country air raid of the war—and the most important raid of the war to date.

A briefing of the crews from Clark was set up for that night. I was waiting for my turn to brief on the rescue procedures that would be followed during the mission should a plane go down. I would be in an airborne Command Post, orbiting over the Mekong River, south of the target area. Colonel Ross, the Director of Operations of the Second Air Division, would be flying the plane I would be in. The weather officer was to brief before me. He told the crews that there would be high, broken clouds at about 15,000 feet and a few scattered clouds at about 7,000 feet, with good visibility. When the major returned to where I was waiting, I said, "Major, how would you compare the weather in the target area tomorrow, with the weather in that area today?"

He said, "It will be about the same."

I said, "Do you know that today I had to call off the search for the pilots by choppers because it was raining so hard they couldn't see?"

He turned to a captain from the weather office and said, "Go check the weather." This, after he had just briefed for the most important air strike of the war. The next mistake took place later that night, when Washington changed the bomb load for the strike. They said the planes were not to carry Napalm, only rockets and bombs. (I think they were afraid of repercussions from China, if they used Napalm.)

The next morning, I boarded the C-123 with Colonel Ross as pilot, and we headed for the orbital point over the Mekong River. I was sitting at a table in the cabin with my maps of the target area, monitoring the strike frequencies. The target was a walled area, about the size of a football field. There were two bamboo buildings in one corner of the enclosure. Two trucks were parked in another corner of the enclosure, and the main target—a three antiaircraft gun emplacement—was in the center of the enclosure. The mission was to destroy all traces of everything in the compound to show "them" *what we could do.*

The sequence of events that followed was beyond comprehension. The target was in a long valley. At one end of the valley was a lake. At the other end was a native village. I was monitoring the radio strike frequency, and I heard our "Glorious Red Leader" say, "This is Red Leader. Let's go back to the lake." After an interval of time, "This is Red Leader. Let's go back to the village." He couldn't find the target. This is one reason I thought the navy pilots should have been given the mission. After a few more trips, he said, "This is the Red Leader. I've got it. I'm rolling in." He rolled in at 11,000 feet and pulled out at 7,000 feet, trying to hit a target the size of a football field. I think the reason he pulled up so high was that he was afraid of being shot at. After several passes, they left the target.

The next transmission I heard was from Red Leader's wingman, who said, "Red Leader, you still have external ordinance."

Red Leader said, "How do I get rid of it?"

His wingman said something like, "Do you see that switch on the left panel? You turn that on." It sounded like an instructor telling a cadet how to operate the controls on his first flight.

After the strike, the planes were to hook up to a tanker aircraft to refuel before returning to Danang. The next transmission I heard was from the tanker. He said, "Red Leader, try again. Red Leader, try again." After two or three more futile attempts, the tanker pilot said, "Red Leader, break away. We have other chicks to feed." Then Colonel Ross asked our glorious Red Leader what his intentions were. He said that before proceeding to Danang, he was going to land at Ubon, Thailand, near there for fuel. We followed Colonel Lavin and his wingman to Ubon. When landing, Red Leader ran through the runway lights, blowing out one of his main gear tires. Colonel Ross radioed Danang and asked them to fly a tire to Ubon for Colonel Lavin's jet.

After we landed and Colonel Ross was talking to Colonel Lavin, I got the captain, who was Colonel Lavin's wingman, off to one side and said, "Captain, what in the hell was going on up there?"

He said, "What do you mean?"

I said, "You know what I mean: The colonel leaving the target without expending all his bombs, not knowing how to release them, and then not knowing how to hook up to the tanker."

He said, "Well, the colonel doesn't know his equipment." Yet he was leading the important mission. They brought the tire from Danang, and Lavin and his wingman took off for Danang, arriving there after dark, running through the runway lights and tearing up both main gear tires.

The next morning, I went to Division to see the post-strike photos that had been taken of the target. It looked like two rockets had hit the wall of the compound. One of the two buildings had minor damage, and there was minor damage to one of the trucks. If you stretched your imagination, you could think maybe, just maybe, one of the three guns was damaged. Colonel Ross was there, and he said, "We blew that one."

I said, "No, we didn't. The mission was a complete success."

He asked, "How do you figure that?"

I said, "The mission was to show them what we could do. We showed them we could do nothing. We were a paper tiger."

# Chapter XXXIX

## THE NEED FOR A PLAN

When I got to Vietnam, there was no plan for conducting a rescue mission, even though my predecessors had been there a year. I decided it was high time we had one. With the help of my controllers, we put together a comprehensive plan, which was written along the lines of the way I conducted the foregoing missions. When it was completed, I sent it to the Thirteenth Air Force at Clark AFB. They were over the Second Air Division. They forwarded it on to Pacific Air Rescue Center (PARC) in Hawaii. A week or so later, I was awakened about midnight by the gate bell. I pulled on my pants and went to the gate. It was the Air Police. They said I was wanted at Division. At Division, I was told to call Clark AFB. A major answered and said they had gotten my plan back from PARC in Hawaii, and PARC had made a lot of changes to it. He said he was leaving shortly for Saigon and asked if I could rewrite the plan and put it back the way I had it. I told him I could and would wait for his arrival.

When he arrived, he showed me the plan, and I couldn't believe what they had done to it. I didn't know who the idiot was who had changed the plan, but whoever he was, he had no concept of what a mission was like in Vietnam.

The SA-16, a plane that can land on land or water, is used on large rivers, lakes, and the oceans. Their plan was to land the SA-16 on rivers in Vietnam to rescue pilots that had bailed out. The odds of a pilot bailing out near a river in Vietnam were great enough, but the odds of being able to land on a river in Vietnam, close enough to pick him up, were even greater.

The revised plan said that chopper's going to a crash site would fly at a high speed on an erratic course. Whoever wrote that had been watching too many war movies, in which the soldiers ran zigzag to minimize the chance of being hit. That just isn't the way you do it in a chopper. Captain O'Connell, one of my controllers who was helicopter-qualified said, "If you fly one of rescue's choppers at a high speed, it will fly an erratic course."

There were many more changes to the plan. I spent the rest of the night and all of the next morning rewriting the plan. I took the major to the flight line for his return to Clark. As he got on the plane, he said, "We won't tell PARC about this."

# Chapter XL

## UNRESOLVED PROBLEMS

I had a lot of administrative problems in Detachment 3 which, for the most part, were caused by the administrative section of PARC and the personnel section of Rescue Headquarters in Orlando, Florida. I wrote a letter to Colonel Derck, my commander, recommending that I go to PARC in Hawaii to discuss my problems. I received a message approving the trip. I had made a list of things to discuss with him that covered three legal-pad pages. I spent three days sitting in the orderly room, waiting to see "His Royal Highness" before I was given the opportunity. About 9:30 of the third morning, I was told that Colonel Derck would see me. I went into his "sanctuary" and started down my list of problems, while noticing that he kept looking at his watch. Finally, he said, "Do you have anything real urgent?"

I said, "Colonel, I have a lot of problems to discuss."

He said, "We'll do it tomorrow. I have an important meeting I have to attend."

I said, "You're the boss, but I have some real problems that have to be resolved" (the problems being important enough that I had flown 8,000 miles to discuss them, only to cool my heels three days as I waited to see him). He got up, put on his cap, and walked out the door.

About 11:00 a.m., a message came from my office in Saigon. It said that Captain Cox had been taken to the hospital. Captain O'Connell was on emergency leave in the States, which left my unit without an officer. I had to return to Saigon on the first available plane. Until I got there, Vietnam would be without a rescue controller. In the interim, one of my sergeants would have to handle it till I got there and pray there were no missions.

I asked where Colonel Derck had gone and was told he had gone home. I thought that that was strange since he had said he had an important meeting. I got his home phone number and called. Mrs. Derck answered the phone, and I asked her if Colonel Derck was there. She said, "No, he had a golf date this morning, and he hasn't come back from the golf course." (This was the "important meeting" he said he had to attend.) I told Mrs. Derck about the message and that I had to return to Vietnam as soon as possible. I was taught that the mission was everything. I guess Colonel Derck thought that golf was.

I went to the terminal and saw the passenger service officer and told him my problem. He said he would get me on the next plane, even if he had to bump someone. The next flight was due to depart at 8:00 p.m. I went to my room, packed my things, and returned to the terminal. Shortly before boarding time, I was sitting in the terminal, and Mrs. Derck walked up. She said she was sorry I had to leave so soon without going to the beach or doing anything and that Colonel Derck had come down to see a general off. (Maybe the general was his golf date. If the general stopped quickly, it would probably have broken Colonel Derck's neck.) He never came over to tell me good-bye or ask about my problems. He couldn't have cared less about what was going on in Vietnam. Mrs. Derck was nice enough to say she hoped I found everything "okay" when I got back to Vietnam. So much for my problems, many of which went unresolved.

# Chapter XLI

## BRIEFING THE SHOPPERS

I received a message that General Williams, Commander of Rescue Service in Orlando, and his staff were to be in Clark AFB in two days, and I was to go there and brief them on the situation in Vietnam. (They were on a shopping trip to Hong Kong and couldn't be bothered with the war at that time.) The next day, I caught a flight to Clark, which had a rescue unit similar to mine, except they didn't have a war.

Before I left Saigon, Colonel Allison Brooks Deputy Commander of Second Air Division, asked me if I would take a message to Colonel Tatum, Director of Operations of Rescue. I said, "Sure."

He said, "Tell him I said to get his ass over here and see what you need and give you some help."

I said, "I'll be glad to deliver your message, Colonel." He said that I had just been awarded another Bronze Star, and he thought it would be nice if General Williams presented it to me. I flew to Clark, and when the plane came in from Orlando, we went out to meet the general and his staff. It was obvious to me and others that Colonel Derck snubbed me when he got off the plane. He didn't even speak or acknowledge my presence. The general and his staff shook hands with me and the others in the reception party.

*Receiving the second Bronze Star for Heroism, presented by Colonel Tatum, Director of Operations, Air Rescue Service, for the first mission to VC Valley.*

# THE UNITED STATES OF AMERICA

TO ALL WHO SHALL SEE THESE PRESENTS, GREETING:

THIS IS TO CERTIFY THAT
THE PRESIDENT OF THE UNITED STATES OF AMERICA
AUTHORIZED BY EXECUTIVE ORDER, AUGUST 24, 1962
HAS AWARDED

## THE BRONZE STAR MEDAL
(FIRST OAK LEAF CLUSTER)
TO

MAJOR ALAN W. SAUNDERS, AO169702
UNITED STATES AIR FORCE

FOR

HEROISM

While serving with friendly foreign forces engaged in
an armed conflict against an opposing armed force.

GIVEN UNDER MY HAND IN THE CITY OF WASHINGTON
THIS        24TH        DAY OF        FEBRUARY        1964

LIEUTENANT GENERAL, USAF

*Award of the second Bronze Star for Heroism.*

Major Lane, the Detachment Commander at Clark, set up a briefing at the Officers Club the next day. I briefed the general and his staff on some of the missions I had conducted and on the situation in Vietnam. After the briefing, Colonel Tatum, not General Williams, presented me with the Bronze Star medal, with the V device for Valor.

After the presentation, we went to the bar for drinks. Colonel P.Y. Williams, Director of Personnel for Rescue, came over to congratulate me on receiving the medal. He said, "If it had been in wartime, that would have been a 'Silver Star.'"

I smiled and thought, 'How stupid.' It just showed how little he knew about the situation in Vietnam. He may not have thought it was a war, but I sure as hell did. Every time I went on a mission, we were under fire. Colonel Derck didn't even come over and congratulate me or ask about my problems. I gave Colonel Tatum the message from Colonel Brooks.

That evening, I was having dinner with one of the controllers, his wife, and son. We were discussing rescue in general, and I was telling them about Vietnam. The captain's wife said, "The next time General Williams comes to Clark, all the rescue wives are going to Baggio up in the mountains. The general thinks the rescue wives are his personal harem."

The captain's son was setting off some kind of fireworks called watusis. They look like sticks about the size of the head of a kitchen match and are about

6 inches long. He said if you put them on the floor or sidewalk and someone stepped on them, they would go off with a bang. Also, fire would explode them. I decided I had a use for some of them. The boy said he would give me some.

The Combat Operations where I worked had partitions separating the offices. They were about 7 feet high and about a foot off the floor. Some of the offices were used by Americans, and others were used by Vietnamese. Some of the Vietnamese had a bad habit of going in the American areas and stealing cigarettes that were left on desks or filing cabinets. They even reached under the panel one night and tried to take a pair of boots from rescue.

The stealing had to stop, and I had a plan to catch one of the thieves. When I got back to Saigon, I borrowed a pack of cigarettes from one of my men, as I didn't smoke. I took three or four cigarettes out of the pack, to make it looked

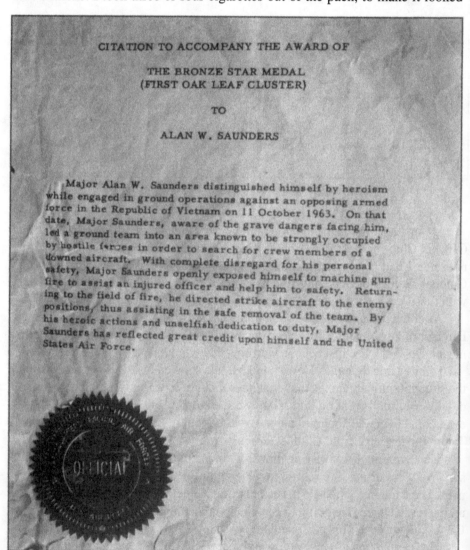

CITATION TO ACCOMPANY THE AWARD OF

THE BRONZE STAR MEDAL
(FIRST OAK LEAF CLUSTER)

TO

ALAN W. SAUNDERS

Major Alan W. Saunders distinguished himself by heroism while engaged in ground operations against an opposing armed force in the Republic of Vietnam on 11 October 1963. On that date, Major Saunders, aware of the grave dangers facing him, led a ground team into an area known to be strongly occupied by hostile forces in order to search for crew members of a downed aircraft. With complete disregard for his personal safety, Major Saunders openly exposed himself to machine gun fire to assist an injured officer and help him to safety. Returning to the field of fire, he directed strike aircraft to the enemy positions, thus assisting in the safe removal of the team. By his heroic actions and unselfish dedication to duty, Major Saunders has reflected great credit upon himself and the United States Air Force.

*Citation to second Bronze Star.*

174

used. I took a short piece of watusi and put it down inside the end of the next four cigarettes. I placed the pack on the file cabinet near the door that night. I told the controller to come with me, and we went into my office and waited. We didn't have to wait long. There was a loud bang in the Vietnamese section across from our section. We went over and looked in. There was a Vietnamese Airman standing there, and his face was black from the explosion. A Vietnamese guard who heard the noise came in and wanted to know what had happened. I said, "I don't know. Why don't you ask that man?" I pointed to the airman with the black face and powder burns on his hand. The guard took him away, and he never came back. The word spread, and the cigarette stealing stopped.

# Chapter XLII

## VIP VISITS AND LIES

Saigon had many visits by VIPs: Secretary of Defense McNamara, Secretary of State Rusk, many Congressmen, and others. You could always tell when a VIP was coming to town. The Vietnamese government would put up welcoming banners across the main thoroughfare from the air base to downtown Saigon, especially if the VIP could provide them with more money.

Briefings would be set up by the American Military Command, the Vietnamese Government, and the U.S. Embassy. MAC-V (Military Assistance Command, Vietnam), as well as the American ambassador, had put out the word that the name of the game was, "We are winning," and any unit commander who said anything to the contrary would be relieved. If the VIPs had asked anyone below the rank of colonel, they would have been told that we were losing our shirts and the war. The generals were so afraid of losing their jobs, they wouldn't say, "Mr. Secretary, we're losing our shirts over here, and I can't stop it"—which was the truth. They were afraid that Mr. McNamara would say, "Maybe we need a new commander over here. I'll try to find you a job in the States." So MAC-V and the ambassador covered up the truth, and the VIPs would go back to Washington and tell everyone that we were winning the war.

The Vietnamese set up itineraries for the visiting wheels to visit two or three hamlets, an enclosed area that had been set up to protect the villagers from being overrun by the Vietcong. It was all a sham. Many of the hamlets had been overrun anyway. They were called "Strategic Hamlets," "New Life Hamlets," or some other name. Every time the program failed, they would start it up again with a new name.

The Vietnamese officials would pick out a hamlet that had not been overrun, and the villagers were told to clean it up and get it ready for the visitors. Before the time for the visit, Vietnamese government employees from Saigon, dressed like villagers, would be flown to the hamlet from Saigon. They were told what to say when answering questions from VIPs. They would be scattered throughout the crowd, and the local natives were told to keep their mouths shut. The "stooges" from Saigon would extol the many improvements that had been made with the aid received from the USAID. Then the VIPs would be taken to a rundown hamlet, usually one that had been overrun by the Vietcong. The VIPs would be told by the escorting officials, "If we had more money, we could make this hamlet look like the other one." They always had their hand out for money, and they always got what they wanted. The U.S. Embassy and USAID made many of the government officials rich. I think the greatest fear of our State Department officials was that they would make some country so mad that it wouldn't take our aid, so they gave it anything it wanted.

A Vietnamese military officer told me that what surprised him most was how gullible our statesmen were. He said, "We take them where we want them to go, show them what we want them to see, and tell them what we want them to hear, and they go back to Washington, thinking they know what is going on over here."

In 1963, Ambassador Ellisworth Bunker was a loser as ambassador to Vietnam. General Paul Harkins was Commanding General of MAC-V, and Major General Timmes was Commanding General of MAAG (Military Assistance Advisory Group) for Vietnam. On November 1, 1963, the air edition of the *Pacific Stars and Stripes* had pictures of Harkins and Timmes, with a lead story of 1-inch headlines, quoting the two generals. Excerpts from the story follow:

## "VIET VICTORY NEAR"
## 1,000 TO LEAVE SOON – HARKINS

The top two military leaders in Vietnam said that "victory in the sense it would apply to this kind of war" is just months away, and the reduction of American Advisors can begin at any time now.

General Paul Harkins, overall Commander of the 15,000 American troops in this country, told *Stars and Stripes* that "about 1,000 troops will be gone from Vietnam by the end of this year" and will not be replaced.

Harkins' Personnel Chief said detailed reduction plans had been drawn up, and approval from the Pacific Commander-in-Chief in Hawaii is expected "within a few days."

Defense Secretary Robert S. McNamara said earlier this month in Washington that it might be possible soon to start cutting U.S. military strength here. It is expected that the bulk of the first strength cut will come from the approximate 60 percent which make up the rear echelon administrative and logistic element.

In separate interviews, both Harkins, who is Commanding General, U.S. Forces Military Assistance Command, and Maj. Gen. Charles J. Timmes, Chief of the U.S. Military Assistance Advisory Group, painted a highly optimistic picture of the military effort here.

\*\*\*

You can't blame McNamara for believing what his commanders in the field told him. He had to accept what they told him as "fact." I fault him for not relieving them of their command and asking them to resign when he realized he had been duped. It was clear that they were incompetent to command in a theater of war. The generals only miscalculated the end of the war by almost ten years.

Their pronouncement reminded me of the time I was flying the Hump in 1943, and the commanding general sent a message to all the bases, saying, "I note with increasing alarm the drop in tonnage of supplies to China. Let's get on the ball and get the boys home by Christmas." They were still flying supplies to China over the Hump four years later until Chiang was thrown out of China by the Communists.

General Westmoreland, who replaced Harkins, wasn't any better. He conducted an "air-conditioned war." The MAC-V headquarters building was air-conditioned. The wheels lived in air-conditioned villas, a lot of the men and most of the officers lived in air-conditioned hotels rented by the U.S. Government, and a lot of them rode to work in air-conditioned cars. When I first got to Tan Son Nhut, we were living in tents. Later, I thought that if they would put everyone in tents and disconnect all the air conditioners, they would find a way to end the war. They were drawing overseas pay, and a lot of them were getting combat pay. Many were living or sleeping with Vietnamese girls and had never had it so good. (I'm speaking mainly about the air force and the army "Saigon warriors.") I think if MAC-V had put out a notice to the effect that anyone wanting to return to the States could go but could not return, they wouldn't have had to add one extra plane to the schedule.

When the men would get orders to return to the States, their girlfriends would accompany them to the airport, crying because their loved one was leaving. It got so bad, MAC-V put out a notice that the "tearful farewells at the airport had to cease."

Everyone had a DEROS (date eligible to return from overseas); however, in Saigon it was known as "The date you will leave your loved one and return to your wife." Some of the men volunteered to return to Vietnam two or three times. When the Americans finally went home, they left thousands of "Amerasian" babies in Vietnam. I thought that the worst torture some of the air force pilots who were POW's had was wondering who was sleeping with the girl they left the morning they made their last flight. It was funny because as soon as the men would leave the girls crying at the airport and then board the plane for home, the girls would be looking at new arrivals.

# Chapter XLIII

## FRIDAY FOLLIES

Every Friday the news people and war correspondents would gather for a briefing by the MAC-V Information Officer. This day was given the name "Friday Follies" by some because of the lack of valid information. The military used a play on words—specifically the term "lost to combat"—when providing statistics to the news media. Of course, everyone thought any losses in Vietnam, either men or planes, were losses in combat. Not so. "Lost to Combat'" meant that "the bodies or planes and choppers had holes in them." They never mentioned that there was another list of men and planes and choppers that were not "lost to combat." Whenever I would come back from a mission, they always asked if there were bullet holes in the plane, chopper, or bodies. The bodies were usually just pieces of flesh, and I usually didn't hang around looking for holes in the planes. When you are being shot at, you just don't go poking around looking for holes in what's left of the plane or chopper. So the reporters would report what they were told. For example, the briefing officer would say, "The past week, there were twenty American men, one T-28 and two choppers lost to combat. Also, there were 100 Allies and 1,000 enemy lost to combat." (They always reported huge enemy losses.) They wanted to make the ARVN troops look good. They were obsessed with "body counts." The troops were always told to count the bodies of the enemy after an engagement. Then MAC-V would inflate those figures, saying that the enemy had carried the other bodies off with them. I think if you could have counted all of the enemy we reported we killed, we would have been charged with genocide.

When the college students and others started protesting the war and losses in Vietnam, Washington sent word to "hold down the losses." There are two ways you can "hold down the losses": reduce the fighting or stop reporting all the losses. You can't stop fighting, or you lose, so you stop reporting all the losses. When the American public read the figures released of the American losses, they thought, 'That's not bad for a war.'

The way MAC-V rationalized the "not lost to combat" list was, if I didn't see bullet holes in the plane or chopper, "it probably had engine failure and crashed"; therefore, it wasn't lost to combat.

The American people never knew how many American men were lost. If you had a loved one reported killed in action, when twenty men were reported "lost to combat," you would think that your loved one was one of the twenty; however, there may be a hundred like you who thought the same thing. The public never knew the true losses until the Memorial Wall was built with the names of all those lost in Vietnam. The government had to put all the names on the wall because the friends and relatives would be looking for the names of their loved ones.

# Chapter XLIV

## KOREAN INVOLVEMENT

Korea was asked by the American government to send troops to Vietnam. Under the agreement, the American government would pay and equip them. At one time, there were over 50,000 Korean troops in Vietnam. The American military had boasted that the Korean troops were as good as any troops anywhere in the world. (I think they said that because Americans had trained them.) If that was true, why didn't they leave the Korean troops in Korea to protect their country and bring the approximately 35,000 American troops stationed in Korea to Vietnam. Then we could say that there was one place in the world where we went to fight and had removed all of our troops afterward. I can't think of any place in the world where we went to fight and, afterward, pulled out *all* of our forces—except Vietnam, and we were kicked out of there.

I was told that the American Government was paying the Korean troops the same pay scale as the American men with the same rank, plus per diem, which meant they got more pay than the Americans with the same rank. They were also given Post Exchange (PX) and Commissary privileges, which they abused terribly. They would go to the Post Exchange in Saigon, buy TVs and other electronic items that would sell well on the black market, take them out to the parking lot, and put them in a taxi. The taxi would leave with the items, and the Koreans would go back into the PX for more. They would do the same in the commissary. They were selling the goods on the black market, and the American military refused to do anything about it. The military police were told to look the other way and not apprehend them or say anything to them. If an American had done what the Koreans were doing, he would have been court-martialed.

The military paid the Americans and Koreans with Military Payment Certificates (MPC). You were not allowed to have American money. Each issue of MPC was all one color. About once a year, without prior notice, the military would change the MPC to an issue of a different color. They would close the military installations, and you couldn't leave until you had turned in the MPC you had in exchange for the new issue. However, the military gave the Koreans one week to turn their MPC in. All the Vietnamese who had MPC got stuck with it. The Koreans would go around and give the Vietnamese ten cents on the dollar for the old issue. Then they would bring it in for exchange up to a week later. They would bring in as much as a footlocker full, and the military wouldn't say or do anything about it. I think they were a lot like our State Department. The State Department's greatest fear is that they will make some country so mad, it won't take our aid.

# Chapter XLV

## CARRIER TRIPS

The navy had a carrier task force offshore from Danang. Their planes were the ones doing reconnaissance flights for MAC-V. They had a liaison office in Saigon, headed by a navy captain who contacted me and asked me if I could go out to the carrier and brief the aircrews on jungle survival and rescue procedures. I told him I would be happy to. He said he would advise the carrier, and they would fly me out to the carrier in the afternoon. I could brief the aircrews that evening, spend the night, and fly back the next morning. I had never been on a carrier before and looked forward to the experience.

The next day, the captain called and said the plane would be at Tan Son Nhut at two o'clock to take me to the carrier. They sent one of the largest bombers (I think it was an A-3) operating off the carrier. The carrier was the *Constellation*, one of the largest, if not the largest, in the fleet. Upon arrival, I was given a complete tour of the carrier. It was an unforgettable experience .

The operations section told me the admiral would like to see me. I was taken to his office, and he wanted to know about my work and wanted to thank me for coming to brief the aircrews. I had brought him a souvenir from Vietnam, a native crossbow, and some arrows. A short time before, the Vietcong had put an explosive charge underwater on the hull of a jeep carrier in the harbor in Saigon, and it sank the carrier to the bottom of the harbor. In the early stages of the war, the natives would shoot crossbow arrows at the low-flying choppers. I told the Admiral, "Admiral, this is what the Vietcong are shooting at our choppers, but don't underestimate them: They sank one of your carriers."

The admiral said, "By God, they didn't sink one of my carriers." The carrier they sank was a small jeep-carrying carrier, and his carriers were huge. I was just joking with him.

After a lengthy discussion on my rescue tactics and the war, the admiral invited me to have dinner with him and his staff. We had a wonderful dinner of wine and steaks, served on fine china, with beautiful sterling silverware. It was strictly "first cabin."

That evening, all of the aircrews gathered in the briefing room for my briefing. After the briefing, we had a question-and-answer session. They were completely isolated from the action in Vietnam and were interested in knowing what was going on. Afterwards, I was presented with a navy "survival" knife on which they had carved "Honorary Roadrunner" (they called themselves "roadrunners"), which made me quite proud.

The next morning, I was scheduled to return to Saigon. We had to get in the plane about thirty minutes before time for us to launch. It was hot, and I was sweating profusely. When it came time for us to launch (they had two catapults in operation launching the planes), the pilot taxied into position, and the deck crew (a deck hand signals the pilot to start engines) hooked us up to the catapult. When they fired the catapult, we went from a standstill to flying speed in just seconds. Sweat flew into my eyes, and I could hardly see. I don't know how the pilot could see the instruments.

Every month or two, another carrier would come on station and relieve the one that was there. When the *Ticondaroga* replaced the *Constellation*, the admiral on the *Constellation* told the admiral on the *Ticondaroga* about me and suggested that I come out and brief the Ticon crews. The captain in the liaison office in Saigon again asked me to go out and brief the pilots. The *Ticondaroga* was a smaller carrier than the *Constellation*.

When I got out of the plane on the *Constellation*, I was met by two operations officers who took me to my room so I could leave my bag and then took me on a tour of the carrier. By contrast, when I got off the plane on the *Ticondaroga*, no one met me, and I had to have the pilot show me the way to operations. There was a strained air on the ship, not friendly like on the *Constellation*. I was not taken to see the admiral that day. I ate with the pilots and, afterwards, briefed them. The next morning, the executive officer told me I would see the admiral for "three minutes," and he took me to him. I gave the admiral a crossbow I had brought him and briefly told him of my work. He said he wished he had known I was on board, as he would like to have talked to me about Vietnam. It seemed strange that he had been unaware that I was on board. Later, I mentioned it to the captain in the liaison office in Saigon, and he said it was because of the executive officer, who tried to keep everyone away from the admiral.

Later, after we had hooked up to the catapult, the deck hand noticed a fuel leak in one of the engines. We were unhooked and towed out of the way. The launch was completed, and the ship was secured. I was told there would be no more launches that day, so I went into operations and told the operations officer that I needed to get back to Saigon. He said that the only plane taking off that afternoon was a courier going to the Philippines and that I could go on it. I said, "How am I supposed to get from the Philippines to Saigon?"

He said, "I'll call the Philippines and tell them to have an A-3 crew standby to take you to Saigon." The flight from the carrier to Saigon takes less than an hour. The flight to the Philippines took four hours, and the flight back to Saigon took another hour. That's doing it the hard way.

# Chapter XLVI

# PX AND COMMISSARY: WOMEN ONLY

When American GIs stationed elsewhere came to Saigon, they would head for the Cholon Post Exchange to stock up on things they needed. One day, a GI from the Boondocks went to the PX. One thing he needed was a pair of shoelaces. The PX had none but had a large supply of women's hairspray and other items for women. He was unhappy, to say the least. When he went back to his outfit, he wrote his Congressman. He wanted to know why the PX could import all the items for the women and not have any shoelaces for the men. As is the custom, when a Congressman gets an inquiry from one of his constituents, he takes action. If the inquiry involves someone in the military, he contacts the Secretary of whichever service is concerned and requests an investigation into the case. In this case, the Secretary of the Army sent MAC-V an inquiry requesting a reply as to action taken to correct the problem. At each level, the rhetoric gets stronger. By the time the request got to MAC-V, it demanded strong action. The result was that the manager of the PX, a civilian, was fired, and all ladies' items in the PX were put "off limits" to men. They could no longer buy hairspray for their "live-in" girlfriends. Only women working for the American Government and wives of men stationed in Vietnam could buy items for women. This directive applied to all PXs in Vietnam.

Just inside the door to the PX at Tan Son Nhut was a rack of shelves. Above it was the sign "Feature of the Week," referring to the items on the shelves that they were featuring that week. One day, a friend and I went to the PX to see what the feature of the week was. One of the items was a box of douche powder. My friend took one and said, "Watch this." He went over to the ladies' counter, pointed to a can of hairspray, and said, "I want one of those."

The lady clerk said, "I'm sorry, Sir. I can't let you have one of those. They are for women only."

My friend set the box of douche powder on the counter and said, "I thought this was, too." He left the box on the counter and walked away. He had made his point.

## COMMISSARY RIP-OFF

In the Cholon District of Saigon was a large commissary run by the military for the military men and those government civilians whose contract permitted them to use it. It had a number of checkout lanes, with Vietnamese girls as cashiers. The cashiers had quite a scam going: They would select a likely recruit, and when he came to check out, the cashier would discount two or three large items. As an example, if the man had a large ham, which was marked $15, when the cashier moved it by, she would ring up $1.50 and then watch the individual for his or her reaction. If the individual noticed the discrepancy and said nothing,

**183**

the cashier would know she had an accomplice in the scam. The cashiers would talk to one another in Vietnamese and spread the word when they had gotten a new recruit. They would also keep a lookout for anyone who appeared to be watching them. After two or three times, they would recognize the individuals, and then it became "big time." Suppose you had a basketful of groceries and went to checkout. The cashiers would always hit some keys on the register, just in case someone was watching. However, she may ring up 10¢ for a $5 item. You knew you had about $30 worth of items. When she finished ringing up the items, she would hit the total button, and it would show $10. She would then say, "That will be $15." You would give her $15. You saved about $15, and she put $10 in the register and $5 in her pocket. When you multiply that figure by the thousands of customers a month, you begin to get the magnitude of the rip-off. The girls (and maybe others) were doing quite well for themselves. I always thought that the commissary sergeant and commissary officer were in on it. They had to be because any kind of inventory that showed stock received, compared to sales and stock on hand, would reveal the scam—and the books really had to be doctored to cover it up.

# Chapter XLVII

## USAID

The USAID (United States Agency for International Development) was the "giveaway program" for the State Department, in countries all around the globe. I have said before that I thought our State Department's greatest fear was that one of their representatives in some country would do something that made that country so mad that they wouldn't take our aid.

The USAID symbol was two hands clasped together in friendship. They put stickers with this symbol on them on everything they gave Vietnam. USAID had "Prov-Reps" (province representatives) in each province, whose job was to see what was wanted or what we could give them. Whatever the officials in the province wanted, the Prov-Rep would order it from the head office in Saigon. The order would be sent to the warehouse at Tan Son Nhut, and an Air America plane would deliver it to the province.

The waste was tremendous. For example, shipments of cement and grain would be brought from the Port on pallets, offloaded on the ramp near the warehouse, and (for the most part) just sit there until it was flown to the province. It was not covered, and when it rained, the cement would get wet and set up, rendering it useless for making concrete. Some military units on the base would use the hardened sacks of cement to make security bunkers, and the rest would be thrown away. Likewise, the grain would get wet and sour. Some of the corn was so eaten by weevils that only the hulls were left. The area smelled like a Distillery. A friend of mine who was a pilot on Air America told me that on one trip he was supposed to have 10,000 pounds of corn on board. When he took off, the plane literally jumped in the air. He went back and checked the load, and the weevils had eaten all the inside of the kernels.

USAID was, at times, inadvertently supporting the enemy. There was one case where the Prov-Rep ordered rice for a village about 100 miles north of Saigon. There were 110 houses or huts along a road near an airstrip. Air America flew about 60,000 pounds of rice to that village. There were about 500 men, women, and children in the village. The next month, the Prov-Rep ordered another shipment of rice for that village. It doesn't take a Rhodes scholar to figure out that 500 people can't eat 60,000 pounds of rice in a month. What had happened was that the village chief, district chief, and province chief had gotten together and sold the rice to the Vietcong. However, they got the new shipment they requested. USAID didn't want to make them mad, or they might not have wanted more rice. In another case, an Air America chopper pilot told me he had been flying sling loads of 2,000 pounds of cement to a village in the hills in Central Vietnam. He said they were putting the cement under a shelter, and there wasn't an American there. That afternoon, he noticed men with bags of cement on "jog sticks" going over the hill near the village. He went back to the coast and refused to take any more cement to the village until they sent an American there to see what was going on.

As in the case of the rice, the cement was being sold to the enemy by our Allies whom we were trying to help. He said he went back to that village a few weeks later, and there was nothing new made of cement, and all the cement was gone. Some time later, the American troops made a sweep through that area and were hit by machine gun fire from concrete bunkers. There was our American cement. USAID was importing a lot of different kinds of grain for South Vietnamese people because most of their rice and other grain crops had been killed by Agent Orange. One of the grains we imported was bulgar, or coarse ground wheat. The natives didn't like it. They were used to a diet consisting primarily of rice. One day I was at the USAID warehouse where they collected most of the items for shipping to the provinces. An American State Department type with his horned-rim glasses and attaché case came in and told the warehouseman that he wanted to check on his order for cornmeal for the pig program. I said, "You're feeding the pigs cornmeal?"

He said, "Yes."

I said, "Why don't you feed them bulgar? There is lots of it, and the people don't want it. When I lived on a ranch, we put ground wheat in a wooden barrel, covered it with water, let it ferment, and fed it to our pigs. They get fat very quick on it."

He said, "Wouldn't they get intoxicated?" You would expect a question like that from someone from Washington. It pointed out their use of unqualified people in the program.

# Chapter XLVIII

# GULF OF TONKIN RESOLUTION: A BIGGER WAR

The navy had two destroyers go up into the Gulf of Tonkin. The North Vietnamese sent torpedo boats out to attack them, and Washington hollered "foul." They said the attack was unprovoked. Hogwash. The destroyers had been escorting South Vietnamese gunboats up into the Gulf of Tonkin, where they were shelling shore installations. The North Vietnamese just got tired of it and went after the destroyers. As a result of the alleged incident, Congress passed the "Gulf of Tonkin Resolution," which gave the Department of Defense the authority to expand the war and bombing raids into North Vietnam. The justification used to get Congress to pass the resolution was a sham, but it served the purpose: It gave the Department of Defense the excuse it needed to expand the war to North Vietnam.

## ROLLING THUNDER
The air force had moved B-52 bombers into Thailand and Guam and started using them in South Vietnam. The code name "Rolling Thunder" came from the sound made when the 60 or more 500-pound bombs on each plane in a three-plane formation hit the ground. The sound could be heard for many miles, and the shock wave could be felt for a great distance. It had to be terrifying to anyone in the vicinity of the strike.

The B-52 was good for strip bombing where enemy troops were "thought" to be concentrated. (I use the word "thought" because they never really knew because we had lousy Intelligence in Vietnam.) The B-52 bombers were also used on other large targets.

The cost effectiveness of a strike by the B-52 bombers was questionable, at best. I saw one post strike report that said the strike resulted in one bicycle destroyed and a stove uncovered. Compare that with the cost of 180 500-pound bombs, plus the cost of delivery from the factory to the target, and it doesn't seem worthwhile.

The Strategic Air Command (SAC) had touted the B-52 as America's "First Line of Defense." According to the ballyhoo put out by the Department of Defense, if the Russians started anything, the B-52 bombers would flatten Moscow. That is what they wanted the American people to believe, yet when they started using them in Vietnam, they couldn't penetrate North Vietnam airspace because Russian Sam missiles, manned by less experienced gunners, would shoot them down. It wasn't until they developed countermeasures that they could send them deep into North Vietnam, and they still had losses.

They were going to flatten Moscow? Give me a break.

# Chapter XLIX

## STUDY: RESCUE HELICOPTER AND MANNING REQUIREMENTS

When I took over rescue operations in Vietnam, I inherited an office, two desks, two chairs, and one telephone, but I had no transportation, so when there was a crash, I had to hitch a ride to the flight line. We didn't have any helicopters, so I had to beg the army chopper unit for a ride to the crash site. The army choppers were not equipped with a cable and winch needed to pick up a man in the jungle. Nothing had been done by Colonel Derck in Pacific Rescue Center, the staff in Orlando, my predecessor, or anyone else to obtain suitable aircraft, necessary transportation and communications equipment, or trained paramedic personnel.

I decided to prepare a study outlining the requirement for specialized helicopters, equipment, and personnel for rescue operations in Vietnam. The jungle in parts of Vietnam had trees 200 feet tall. That meant that the choppers had to have a wench with at least 250 feet of cable capable of penetrating the jungle canopy and picking up a man on the ground. From my experience in army choppers, I knew it was absolutely necessary to have door guns to suppress enemy ground fire in the vicinity of a crash site. With few exceptions, the plane had been shot down and crashed in enemy-occupied terrain. I included door guns as a requirement.

In addition, the choppers would need to be large enough for a crew of two pilots, two door gunners, a paramedic, and at least six passengers. I considered it a must to have on board a medic who could go down on the cable to give aid to an injured man on the ground and assist him in being lifted back to the chopper. The medic would also have to go to a crash site with the team on the ground, in case there was an injured man at the crash site, and provide medical aid to the injured on the way back to a hospital. I took into consideration the large area to be covered, as well as the stepped-up sortie rate, in determining how many helicopters would be needed and where they would be strategically located.

I submitted the study to Second Air Division for their concurrence and they forwarded it to MAC-V, who was over all military operations in Vietnam. The Joint Chiefs of Staff had given the responsibility of rescue operations to the air force. MAC-V was army, and they didn't agree. They thought they should conduct their own rescue operations, and they sat on the study and wouldn't forward it up the chain of command to Thirteenth Air Force, which was over Second Air Division. They sat on it several months. Meanwhile, we were losing good men for the lack of specialized rescue helicopters, equipment, and para-rescue men.

I was notified that staff members from Rescue Headquarters in Orlando were coming to Clark AFB in the Philippines, and I was to go there to brief them on rescue operations in Vietnam. I decided to take a copy of the study with me to give them an advanced look at it.

I went to Clark, and after the briefing, I was going over the various recommendations I had incorporated into the study. When I told them about the door guns, Colonel Derck said, "The helicopters aren't to be weapons platforms." I told him that I knew from experience that when you were being shot at, if you didn't return the fire, the Vietcong wouldn't even hide. They would stand out in the open and shoot you down.

When I mentioned the urgent need for para-rescue men, Colonel P.Y. Will-

# SECOND AIR DIVISION

# S A R   REQUIREMENTS

# STUDY

# REPUBLIC OF VIETNAM

# 1 SEP 63

*Cover of the study I submitted.*

iams, Director of Personnel, said, "There are no para-rescue men in the Rescue organization."

I said, "Then you need to get some assigned. Don't wait till the modified heliocopters are approved. Requisition them now. There is an urgent need for jump-qualified, medically-trained personnel to care for injured crew members at a crash site and on the way back to a hospital."

Their comments illustrated just how little the staff, including Brigadier Adriel Williams, knew about what was going on in the war in Vietnam or with rescue requirements. I think you could take what they knew, write it on a cigarette paper, tear it off, and roll a cigarette with what was left.

It is significant to note that when the new modified heliocopters arrived in 1965, they had door guns and winches, and para-rescue men had been assigned to Rescue. More knowledgeable minds had prevailed.

In the *Vietnam* magazine, Richard Eichenlub's article "That Others May Live" sums up the result of my study and the effect it had on the operation of rescue in Vietnam. Excerpts from that article follow:

"...They lacked suitable communications equipment, and used messengers on bicycles rather than trust the unreliable Vietnamese telephone network.

There were no rescue helicopters. When an aircraft was reported as overdue, ARS personnel requested available resources from nearby Army or Marine units, and flew to the vicinity. They would search for the wreckage and, if located, remove any bodies. In cases where rescue of survivors was attempted, inadequate training and improper equipment often led to disaster. In one case, two marine helicopters crashed while recovering an OV-10 Bronco observation aircraft pilot on the top of a rocky hill. In another instance, a flier stranded in a river was drowned by the rotor wash when the rescuing crewmen tried to pull him directly into a hovering helicopter.

The Air Rescue Service received the official designation of Detachment 3, Pacific Air Rescue Center on April 1, 1962. When Major Alan Saunders took command of the unit in June 1963, he quickly recognized the woeful situation. Saunders prepared a scathing report requesting equipment specifically engineered for airborne search and rescue operations. He also advocated the creation of a cadre of trained individuals, both fliers and para-rescue men, whose sole responsibility would be the recovery of downed aircrews. Unfortunately, the report stalled when it was routed to the Army dominated U.S. Military Assistance Command Vietnam, (MAC-V) headquarters, for endorsement. Eventually, In May 1964, it was brought to the attention of the Joint Chiefs of Staff, who made the Air Rescue Service part of the Air Force..."

## A SPECIAL SORT OF MAN

"...As the Vietnam conflict progressed, improvements in equipment made Search and Rescue (SAR) more successful. But an even more critical component was the para-rescue specialist, otherwise known as the parajumper, or PJ.

These highly trained individuals provided a variety of services while on a

SAR mission, ultimately being responsible for the successful recovery of the man on the ground. They would fire the primary armament on the Rescue helicopters. This usually consisted of 7.62 mm six barrel mini-guns. In the first helicopters used for rescue missions, the mini-guns fired out the side doors; in later models, they fired out the rear of the rescue aircraft as well. This was the latest available fire capability to keep the enemy forces away during a pickup and to cover the helicopter exit from the area.

More significantly, the PJ would descend to the ground on the wench cable jungle penetrator to locate and assist a wounded airman. Sometimes the injuries were severe enough to necessitate securing the man to a Stokes Litter before hoisting him up. Usually, the PJ would accompany the litter back to the hovering helicopter. Often, this ascent was through considerable hostile fire.

After the recovery, the PJ was also the first to provide trained medical assistance to the rescued airman, and treatment would be applied during the helicopter's return to a friendly air base in Thailand or South Vietnam, or a ship in the Gulf of Tonkin. Their bravery was legendary..."

# Chapter L

# UNTENABLE POSITION

I had been placed in an untenable position, being under the Second Air Division's Operation Control while being assigned to the Pacific Air Rescue Center administratively. My commander, Colonel Derck in PARC, had no concept of what the war in Vietnam was like. For example, Second Air Division asked me to go back to the crash site at Ba To and recover the part of the wing that had broken off, and Colonel Derck sent me a letter criticizing me for doing it—notwithstanding the fact that I was under the operational control of Second Air Division. As previously stated, I would bring out parts of bodies, for which I was criticized by Colonel Derck. On another occasion, a joint exercise was to be conducted by American and Thai forces—in Thailand. I was asked by the Second Air Division Commander what rescue forces were being provided. I told him I was not aware of the exercise and that my commander in Hawaii had not informed me about it or about what forces I was to provide. It was embarrassing to me not to have been told about the exercise by my commander.

Another example of Colonel Derck and his staff's complete lack of understanding of the situation in Vietnam was when I received a letter directing me to conduct a physical fitness test of the men assigned to me. I was to take the test also. We were the laughing stock of the air base. It was probably the only time in military history that men in a war zone had been ordered to take a physical fitness test. I don't think Colonel Derck could have passed a fitness test unless it involved driving and putting a golf ball.

I was due to retire at the end of October of 1965, as the military had a stupid regulation that required you to retire when you had twenty years of active duty, but I wanted to stay in past that time. I requested an extension, which was denied. The military had a saying, "There is no good reason for it; it's just policy"—and that fit my situation exactly. I would be only forty-five when I would have to retire. I was in perfect health and physical condition and had been through three wars—plus there was no way that younger officers could gain the experience I had.

My tour in Vietnam was to end in June of 1964, but I wanted to stay in Vietnam until my retirement because I was doing what I loved to do. I had requested an extension of my tour to my retirement date. The approval from Colonel Derck took a long time to come. (I think he delayed it just to harass me.)

The criticisms and lack of support from Colonel Derck continued. I think the final straw that broke the camel's back was when I received a letter from him, saying that my administration of the Detachment was unacceptable. I think the whole problem between Colonel Derck and me stemmed from the different values we placed on things. He thought that the important things were to have a low handicap, keep a date on the golf course, and make the tee-off time. On the other hand, I thought that the most important thing was to provide rescue for the ever-increasing losses of aircraft and crews in the war. I considered that a higher priority than paperwork or playing golf.

# Chapter LI

## FED UP WITH THE SETUP

I had had it. I wrote a letter to General Williams, Commanding General of Rescue, advising him of the problems I had been having with Colonel Derck and his lack of support. I requested that my unit be given an inspection by the Inspector General's Office of Rescue and that my tour extension be cancelled. I also requested that I be transferred from under Colonel Derck's command.

I received a letter from Colonel Williams, Director of Personnel. He referred to my letter to General Williams and wanted to know if I wanted a formal IG (Inspector General) inspection or a routine inspection. I sent him a message stating that I just wanted an inspection of the administrative functions of the Detachment to show that there was nothing wrong with my administration of the Detachment, as indicated by Colonel Derck. I didn't realize at the time that I was being set up for an inquisition.

After my mother passed away in 1982, I was going through some of her things and found a letter I had written her from Vietnam about the same time that the above was going on. Some excerpts follow:

Dear Mother:

I'm kinda down in the dumps. Guess I'm beginning to let this place get to me. Have been trying ever since I got here to get some rescue aircraft and equipment over here to help save some of the men who are going down. I thought it was all settled for them to come, but now it looks like they are postponed indefinitely. I think the Army is fighting it. You know the Army and Air Force are battling over who does what in the air over here.

Have been considering writing my Congressman, O.C. Fisher, and asking him to ask Defense Secretary McNamara why there are no Rescue aircraft or trained crews in Vietnam. We have lost 135 crewmembers here, and some could have been saved if we had the proper equipment here.

I have been getting no help from my Headquarters, only criticism. My Commander has only been here twice, both times last year. The first time for forty-five minutes, and the next time for two hours. Both times, he was on his way to Hong Kong to go shopping. That was more important to him than helping me with all the problems I have been having. His shopping in Hong Kong had priority over helping to save lives in the war. I have put up with all I can from him and his staff. I have asked to be removed from his Command as soon as possible.

Well, twenty-three years ago yesterday, I joined the Army Air Corps. I will get out in October 1965 with twenty-four years' service. For the most part, they have been good years, with a few not so good. I can look back and see several things I should have done differently and wrong decisions I made, but I guess that could be said of anyone. All in all, I enjoyed it. I considered it a duty and a challenge. Well, so much for unburdening.

## DIRTY POOL

I received a message from Colonel Derck, saying that Colonel P.Y. Williams and a Lieutenant Colonel Maas from Rescue Headquarters were coming to Saigon. I thought that they were coming to make the inspection I had requested. *Wrong.* When they arrived, I was handed a copy of orders relieving me as commander of the detachment, the same order assigned Lieutenant Colonel Maas as commander and me as his assistant. This was "Dirty Pool," and I smelled a "rat"—and the "rat" was Colonel Derck.

## A KANGAROO COURT

The day of his arrival, Colonel Williams ordered me to report to him in the Second Air Division Conference room at 8:00 a.m. the next day.

I reported as ordered and still didn't realize that I was being "set up" and that the colonel had been sent there to conduct an inquisition. When I saw Mrs. Lambert, the Second Air Division court reporter there, and was told by Colonel Williams that I was to be sworn in, I began to realize that this had nothing to do with the administrative inspection I had requested.

When Colonel Williams told me I was to be "sworn in," I told him that if I was required to testify under oath, I wanted to be represented by an attorney.* He said that that was not necessary and proceeded to place me under oath to "tell the truth, the whole truth, and nothing but the truth, so help me God." He said, "Do you understand your rights?"

I said, "I am aware of them."

He said, "You are aware of them?"

I said, "Yes."

Colonel Williams started off by giving me the "third degree" about the letter I had written General Williams and what my reason was in writing the letter. It was obvious from the outset that he had been sent there to denigrate me and to absolve Colonel Derck and his staff, as well as my predecessor and his controllers, of any lack of support or wrongdoing. In addition to telling General Williams about the lack of support from Colonel Derck, I had told him about my predecessor's controllers coming on duty drunk and that no punitive action had been taken against them.

Colonel Williams attempted to refute everything I had told General Williams in my letter. He wanted to know what my purpose was in writing the letter. I explained to him that I wanted General Williams to know what had been going on and that it was Colonel Derck and his staff, as well as my predecessor and his staff, which resulted in the problems with my administration of the Detachment. He asked me to provide him with documentary proof of each of my allegations in the letter to General Williams. Although I was testifying under oath, he wouldn't accept anything I said without documentary proof.

*In the final report, there was no mention of my request to be represented by an attorney. It was obvious that the report had been edited by Colonel Williams prior to the final printing.*

As to my statement about going to Hawaii to discuss the problems I was having with Colonel Derck, as well as having to wait three days to see him, only to have him then leave to play golf, Colonel Williams asked me if I had had an appointment to see Colonel Derck. I said I didn't have an appointment for a specific time to see him, but he knew I was there, and he didn't call me in to discuss my problems until the third day. Then he left for what he said was an "important appointment." I found out later that the "important appointment" was a golf date.

Regarding the statement in my letter to General Williams that one of my predecessors' controllers had been drunk on duty, Colonel Williams asked me if I had seen him drunk on duty. I explained that I had not yet arrived in Vietnam when the controller had allegedly been drunk on duty. I told him that the combat control officer had told me about the incident and that the SAR logbook for that date had the page torn and the scribbled note "There are no damn pencils." I showed him the torn page in the logbook. He questioned my statement in my letter to General Williams that nothing had been done to the controller for having been drunk on duty. He asked if I had seen the controller's efficiency report. I said I hadn't seen it. He asked how I could state that nothing had been done to the controller. I told him that when I asked Major Trexler what he had done to the controller, he said he had done nothing. I asked Colonel Williams to interrogate my controllers—Captain Cox, Captain O'Connel, and S/Sgt. Tedder—about the incident.*

Colonel Williams then asked me if I had ever brought out parts of the aircraft when I was at a crash site. (This was further proof that this was not an inspection of my administrative functions but an attempt to discredit me in everything I had done.) I said that I had on a couple of occasions. (He knew that I had because it had been reported in my Mission Reports.) He said it was not my responsibility and was not in accordance with rescue directives. Then, when I brought out parts of a plane, he asked me under what authority was I operating. I advised him that I was under the Operation Control of the Commander of the Second Air Division, and the Deputy Commander had directed me to do so. He was referring to my Mission to Ba To, when I was sent back to get the wing spar.

After Colonel Williams left to return to Hawaii to report to Colonel Derck, Mrs. Lambert gave me a copy of the report of the hearing. I didn't know at the time (but found out when I read the report) that Colonel Williams had called another witness.

A Captain OBanion, who was a controller for my predecessor, had been sent to the Detachment on temporary duty, about two weeks before Colonel Williams arrived. At the time, I didn't know why. I hadn't requested additional personnel. I realized, after I read the report, that he had been sent there to report the difference in the way I conducted rescue missions and the way it had been done by my

---

*In the final report, there was no mention of my request to have my Controllers interrogated . They never were.

predecessor. This was evidenced by Colonel Williams' questions and Captain Obanion's answers.

**C.W**: So you would, if I may put words in your mouth, so to speak, you would regard the RCC [Rescue Control Center] as being well within its responsibilities as spelled out in this mission? [This is referring to a mission by my predecessor.]

**C.O**: Yes, Sir, with the exception of log entries which I can't explain.

**C.W**: Relative to my last question and your answer, would you say that now, that is embracing the period of the last few months, and including the period of your most recent TDY here, would you say that the present policy of the SAR Center is to overextend its authority or overextend its responsibility in monitoring and participating in rescue activities in this area?

**C.O**: Sir, I don't feel that I am qualified to answer that because, as I stated before, since I have been here, since the 18th, we have not been out in the field or gone out after any pickups or any rescues. Previous to that, the only thing that I have to my knowledge written or read about* is on certain missions, or one mission on a recovery of a wing panel which, of course, did not fall into Search and Rescue's responsibility, which was recovered by personnel of the SAR Center. This type of work, of course, is not within the SAR responsibility as such, in the regulations and even the intent.

The report of the mission he refers to was classified "secret," and he wouldn't have had access to it if Colonel Williams had not sent it to him, which violated regulations pertaining to the distribution of classified documents. The copy of the report that Mrs. Lambert gave me was 50 pages long. She told me that she had been a court reporter on dozens of court cases and that she had heard of kangaroo courts, but this was the first time she had witnessed one.

Lieutenant Colonel Maas told me he wanted me to write a plan for rescue operations in North Vietnam. I said, "Colonel, you don't want me to write that plan. Colonel Derck said that both my administrative and operational procedures were unacceptable."

Lieutenant Colonel Maas responded, "You are not going to the States until you write that plan." I think he knew that there was no one in rescue who knew enough about it to write such a plan. I wrote the plan because of my concern for the pilots who would be flying there, not because I wanted to help the "Dummies" in Rescue because, at that juncture, I didn't think I owed them anything.

I had an intense loathing for everyone in Rescue—from General Adriel Williams on down. I hated them for not supporting me in providing rescue for downed

---

*His statement, "written or read about," told me that he had been briefed before coming to Saigon on TDY and had been corresponding with Colonel Williams about what he would be questioned about at the hearing—and the answers Colonel Williams wanted.*

**DECORATIONS AND AWARDS**

TOP LEFT TO RIGHT: Command pilot wings – Senior pilot wings – Original pilot wings – Distinguished Flying Cross – 3 Bronze Star Medals – Medal from Vietnam Government – Medal from the Republic of China – 2 Air Medals for Flying Hump – 3 Air medals for missions in Vietnam - Air Force Commendation Medal – Numerous Awards, including two Presidential Distinguished Unit Citations.

pilots in Vietnam. After all I had done to improve the image of rescue in Vietnam and to get helicopters equipment and para-rescue men for future rescue of downed pilots, I was dragged through a kangaroo court.

Later, I read that General Williams had been awarded the Legion of Merit for his "Early assessment of rescue requirements in Vietnam." He probably received the medal as a result of his submitting my Second Air Division SAR Requirement Study, dated 1 September 1963, over his signature. It sure as hell wasn't because of *his* "Early assessment of rescue requirements in Vietnam."

I finally received my orders to return to the States. My new assignment was to Orlando AFB Florida. I was to be the base director of operations and training. I had been a director of operations, wing level, now base level, yet Colonel Derck thought my administration of a five-man detachment wasn't good enough.

Rescue Headquarters was located on Orlando AFB. I found out later that rescue tried to get my orders changed so I would be assigned to rescue. I guess they needed someone who knew what the hell was going on in Vietnam. Their attempt failed.

# Chapter LII

# A PARTING TRIBUTE

The day before I boarded a plane for the States, I was handed a copy of a press release by the Public Information office of the Second Air Division. A copy follows:

OFFICE OF INFORMATION                    TSN64-8-8
SECOND AIR DIVISION                       FOR IMMEDIATE RELEASE
The "Jungle King" Goes Home
By MSgt. Robert C. Brown

Major Alan W. Saunders, a former Commander of the Air Rescue Detachment at Tan Son Nhut Airfield, Saigon, departed today, after more than a year of duty in Vietnam, which he and all those who knew him and worked with him will find extremely hard to forget.

The day before he left, he received the second Oak Leaf Cluster to the Bronze Star Medal. It is something of the measure of the man and his achievements that he earned all three Bronze Stars during his tour in Vietnam and that all three were awarded for conspicuous gallantry in the field under fire. He also holds the Distinguished Flying Cross and the Air Medal with four oak leaf clusters as a result of his service in World War II and the Vietnam conflict. His Stateside assignment is uncertain; he has asked to return to Vietnam as soon as possible. He is that kind of a man.

As Commander of Detachment 3, Pacific Air Rescue Center, Major Saunders had direct responsibility for all Air Rescue activity in both Vietnam and Thailand. His staff consisted of three officers and two sergeants. Base personnel at all active air facilities in the two countries were assigned to Air Rescue work as an additional duty, but at best they could only deal with crashes on or in the immediate vicinity of their respective home bases. As Major Saunders put it, "When a bird goes down out in the boondocks, it's our baby."

He is definitely not the desk executive type. He led the most difficult missions in person. A rugged, sun-bronzed man with the look of a true outdoorsman, he wore his jungle boots and camouflage fatigues as if he were used to them, which he was. A PIO representative once asked permission to accompany him on a mission. When he hesitated, the other man hastened to assure him, "I'm not afraid of getting shot at."

"You aren't?" drawled Major Saunders, "Well, by God I am! I'm not questioning your courage," he went on after a moment, "but your physical condition. How long do you figure you'd last in a forced march up a muddy jungle trail on a six to ten percent grade?" The request was politely withdrawn, but the Major was not exaggerating the terrain in which he had to operate. On one mission, one of the Vietnamese soldiers posted as security guard was carried off and killed by a tiger.

It was the human tigers of the Vietcong, however, who gave Major Saunders and the members of his rescue team the most trouble. On one occasion, he and his team made a two mile forced march up a boulder-strewn jungle ravine to remove certain critical parts from a downed bomber. The group was under Vietcong sniper fire almost constantly during the mission. So thickly was the area infested with VC that the helicopter which evacuated them from a point more than two miles from the crash site was fired on both on landing and takeoff.

On another mission, the one for which he received his latest Bronze Star Medal, Major Saunders voluntarily entered an area of dense jungle and bamboo thickets which he knew to be heavily occupied by Vietcong guerrillas. Under constant enemy harassing fire, he pressed the search for survivors of the crash until the mission was completed.

A colorful, indeed almost legendary figure in Air Force circles in Vietnam, Major Saunders was identified by his mottled jungle fatigues and by his personal weapon, which his associates only half jokingly called "the main battery." Whenever possible, he carried into the field a 375 Magnum express rifle, undeterred by the combined 14 pound weight of the weapon and cartridge belt. "Hell," he said, "If I'm going to carry a gun at all, I want one that will do some good. If a tiger charges me, I want something that will knock him kicking the first shot. If a VC so-and-so is shooting at me from behind a palm tree, I want to be able to cut down the palm tree and get him too."

Alan Saunders is not an egotist, but neither is he a shrinking violet. He is not the "Shucks, it wasn't really anything" type. Like most men who are honest with their friends and themselves, he takes pride in a job well done and accepts his honors as a tangible recognition of genuine achievement. The friends and associates who variously named him "The Jungle King" and "The Tiger" (out of his hearing, for the most part) did so in affectionate recognition of the nature of the man.

\*\*\*

They told a joke in Vietnam about a young GI who was sent to Vietnam. He went to his first sergeant and said, "Sergeant, I know we are fighting with the South Vietnamese and against the North Vietnamese, but how do you tell them apart? They are all Vietnamese."

The sergeant said, "That's easy. If you're going down a trail in the jungle and a Vietnamese comes out of the jungle, you shout, 'Ho Chi Minh is a Son-of-a-Bitch!' If he acts belligerent, he's a VC. Shoot him."

A few weeks later, the sergeant saw the young GI, and he was all bandaged up and on crutches. He said, "What happened to you? Did you mistake a North Vietnamese for a South Vietnamese?"

The GI said, "I was going down a jeep road in the jungle, and a Vietnamese came out of the jungle onto the road. I remembered what you said and got my gun ready and said, 'Ho Chi Minh is a Son-of-a-Bitch!' He shouted back, 'Nixon is a bastard!' We were slapping each other on the back and shaking hands, and a jeep ran over both of us."

# Chapter LIII

# BUREAUCRACY

When I went to Vietnam the first time, I fell in love with that country and wanted to return. I decided to apply for a job with the U.S. Agency for International Development. I was told that if they wanted you, they could take you out of the military. I went to Washington to see if I could file an application for a province representative.

I first went to see General Anthis, who was with the Joint Chiefs of Staff. He knew me well from my work in Vietnam when he was Commander of Second Air Division. I told him what I had in mind and said I hated to waste my experience in Vietnam. He said he agreed and would give me a letter of recommendation. I took the letter to USAID Headquarters and filled out an application for the job. They gave me an appointment for an interview the next day. I was interviewed by a panel of four Department Heads. They told me, "You have the kind of qualifications we are looking for. We will pass this on to Saigon."

I was taken in to meet the head of USAID. In the course of our conversation, he asked me, "What can we do to help the people in Vietnam?"

I said, "There are many things that need to be done, but foremost, you need to help the farmers. The farmers plow their fields with buffaloes pulling wooden plows. I can take you to a store in Saigon, where they have large OLIVER tractors and gang plows for sale. They had the 'Hand-Clasp' USAID sticker on them showing you gave them to the Vietnamese government, but only big landowners can afford to buy them. The small farmers continue to plow with a buffalo and wooden plow. Before the buffalo can pull the wooden plow through the ground, the fields have to be flooded to soften the soil. The farmers have devised several ways to flood the fields. One way is to attach paddles on the back wheel of a bicycle, which splashes water from a canal over the dike into the field. In another method, they have a wooden keg with two ropes attached. Two men flip it into the canal and over the dikes into field. These are very primitive methods, and they have to flood thousands of acres before every planting.

"I can take you to another store in Saigon, where the floor is covered with gas-operated water pumps. They, too, have the USAID sticker on them, and they are for sale. Who can buy them? Not the small farmer. Why don't you have your technical representative in each province that has a lot of farming determine how many tractors and water pumps are needed, and provide them to the provinces? They could pull the pumps on sleds behind the tractors, and when the field is plowed, it can be flooded with the pump for planting the rice."

He said, "If we did that, they wouldn't have anything to do." I said to myself, 'Damn. He's just like all the rest of the wheels in the State Department. He can't think down at the "Rice Roots" level.'

I said, "If they just went over and sat down on the dike and watched, they may think, maybe the Americans and my government are trying to help me, and they wouldn't turn to the Vietcong for help. By not helping them, the Vietnamese and American government policies are driving them to the VC For help." USAID never did help the poor farmers—only the rich—and it wasn't until years later that the Japanese helped them with mechanized equipment.

I returned to Orlando, and after a period of time, I phoned USAID to inquire about the status of my application. The man I talked to said they were waiting on a reply from Saigon. He said they were as anxious to get the reply as I was. He

THE JOINT CHIEFS OF STAFF
WASHINGTON, D.C. 20301

OFFICE OF THE SPECIAL ASSISTANT FOR
COUNTERINSURGENCY AND SPECIAL ACTIVITIES

THE JOINT STAFF

17 March 1965

TO WHOM IT MAY CONCERN:

Major Alan W. Saunders was my Search and Rescue Officer while I commanded the 2nd Air Division in Vietnam. In performing this duty, Major Saunders demonstrated a high degree of aggressiveness in efficiently accomplishing his duties. He is the kind of self-starter who is not satisfied with things as he finds them, but makes a determined effort to improve the efficiency of operations. He was constantly in the field finding out about things. Because he was able to analyze the complex search and rescue problems in Vietnam and develop logical solutions to these problems, he was a highly productive and valued member of my immediate staff.

In the routine performance of his duties, Major Saunders was in continuous contact with representatives from the other Services and various departments of the Government of Vietnam. In these contacts he displayed a high degree of diplomacy and understanding of the culture and customs of the host government.

I would be highly pleased to have this man work for me on any assignment requiring tact and diplomacy while, at the same time, getting positive results in the shortest possible time.

ROLLEN H. ANTHIS
Maj General, USAF

*Letter of recommendation.*

203

said they would send a message requesting a reply. About a week later, I received in the mail the file they had on me with a brief note that said Saigon had said I was "overqualified" for the job.

I caught a flight to Washington and went to USAID Headquarters. When I went in to see the Director, he said, "Have a seat. I'll get your file." In a couple of minutes, he came back with a puzzled look on his face and said, "I can't find your file."

I held up the file I had received in the mail and said, "Do you mean this?"

He said, "Where did you get that?" His response told me he didn't know what was going on in his own office . I told him it had come in the mail, and I wanted to know what was meant by my being "overqualified" for the job. All he said was that Saigon had the final approval authority, and they had disapproved my application.

I went back to Orlando and later met a field representative of USAID. I told him what had happened and asked him what being "overqualified" meant. He laughed and said, "What it means is whoever you were going to work for in Vietnam was afraid, because of your high qualifications, you might get his job, so he disapproved your application." So much for State Department Bureaucracy. It's no wonder they did such a lousy job in Vietnam.

# Chapter LIV

## STATESIDE DUTY: THE GOOD LIFE

As Director of Operations at Orlando AFB, I was responsible for providing a plane for the Air Photographic and Charting Service, a tenant on the base. They would send teams all over Central and South America to take pressure readings. The base had a DC-3 that it kept at McCoy AFB near Orlando because Orlando AFB did not have a runway. The Charting Service would call us when they needed to send a team out, and I would schedule a crew to take them. The better trips I would take myself.

I got a call one day, and they said they had to send a team to the Panama Canal. I checked on it and found that the flight had to go to Homestead AFB Florida to get an overseas clearance, then to Jamaica to spend the night. The next day, it would take the team to Panama, leave them, and return to Jamaica. The following afternoon, it would return to Orlando. I thought that would be a great trip, so I scheduled myself as pilot—and a captain who worked for me as copilot. I told the captain that I had read that when they built the Panama Canal, they had a lot of problems with malaria and other diseases, and I thought it would be a good idea if we took a couple of nurses from the hospital with us "just in case we got sick." He said he thought that would be a good idea. I told him I knew the two nurses we should take, and I would get an "okay" from the hospital commander for them to go. I played pool a lot with the hospital commander, and that afternoon, over a pool game, I told the colonel what I wanted to do. He smiled and said he thought they had cleaned up the canal area, but he would let them go anyway. He said they could use some "R&R."

On the scheduled day, we took of off for Jamaica, with a short stop at Homestead AFB for our overseas clearance. We arrived at Jamaica about mid-afternoon, after skirting Cuba, which we were not permitted to fly over. I had gotten the name of a motel in Montego Bay, Jamaica, and had called and made reservations for everyone. That evening, we had a wonderful time on the beach.

I scheduled the flight to Panama for the next morning. We left the nurses and took the team to Panama. We stayed there just long enough to refuel and returned to Jamaica, arriving about mid-afternoon. That afternoon we went down to the port and watched the natives loading banana boats. That evening, we went back to the beach, where we saw some natives doing the "Watusi" dance. They had a pole suspended between two posts in the sand. The natives, one by one, would bend backwards and dance under the pole to the beat of drums. Each time they were successful, the pole would be lowered to the next level. If they knocked the pole off, they were disqualified. The last one to go under the pole lower than anyone else won. It was amazing how low they could bend over backwards and go under the pole.

I told everyone they could sleep in the next morning and we would take off after lunch and return to Orlando. We had a wonderful time. The best part was that we had to return thirty days later to get the team.

The best trip to come up was a trip to take a team to South America. The team was to take readings all the way down the East Coast and back. The trip was to take thirty days. I scheduled myself as pilot. Alas, it was not to be. The base commander got word that the base was to be transferred to the navy, and he

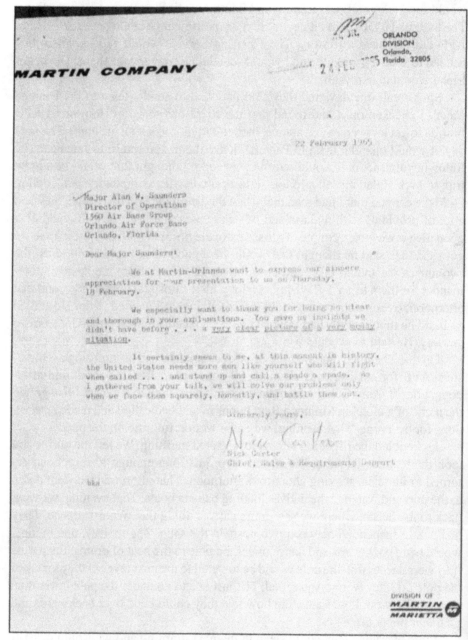

*Letter of appreciation from the Martin Company.*

made me the project officer. I had to take my name off the trip to South America. I sent the captain who was with me in Jamaica, and he was gone a month. What a *rough* assignment.

Civic Organizations frequently look to the military bases for programs or guest speakers. The bases are happy to accommodate them because it is good for public relations.

Shortly after I arrived at Orlando AFB, the base public relations officer asked me if I would be a guest speaker at one of the civic organizations in Orlando. I had always enjoyed public speaking, so I accepted, and it was the beginning of many requests for me to speak. While at Orlando, I gave over forty speeches at civic organizations, universities, and the Martin Company. I was also a regular speaker at the NCO [Non Commissioned Officer] Academy in Orlando, in their "Counter Insurgency" class, as well as a guest speaker on the TV show *Florida Showcase*, which was patterned after the program *Meet the Press*.

I had collected a lot of artifacts overseas, and I gave three two-hour programs on educational TV, which was monitored in the schools both in and around Orlando. I displayed the artifacts, described them, and told where I had obtained each one. It was like a travelogue.

In my talks about the situation in Vietnam, I didn't paint a very good picture of the way the war was going or being conducted. I was predicting that we would have 100,000 men in Vietnam in 1965 and 250,000 to 500,000 in 1966—or we wouldn't be there. Meanwhile, Secretary of Defense McNamara was publicly saying we would have the "boys" home in 1965.

I gave a talk to an organization of retired admirals and generals who met at the Orlando Officers Club. I was seated at the head table, between an admiral and a general. After we finished our meal, the general introduced me. After my talk, I always took questions from the audience. Afterwards, I went back to my seat. The general said, "Does Washington know you are talking that way?"

I said that I didn't know if they did and that my guidelines were that I could talk about anything I had knowledge of.

He said, "You will never make Chief of the Air Staff, talking like that."

The admiral leaned over and said, "That would be a shame, wouldn't it?"

In February of 1965, I was invited to be the guest speaker at a meeting of the Orlando Division of the Martin Company of Marietta, Georgia, a Defense contractor. After a wonderful meal, I was introduced by Mr. Nick Carter. In my talks, I always tried to give as complete a picture as I could of what was going on with the U.S. involvement in the war and the problems (most of which were of our own making) we were having in controlling the situation. Later, I received a very nice letter from Mr. Carter. (See copy.)

The base commander, Colonel Brown, presented me with two air medals that I had been awarded for missions in Vietnam. At my retirement, he presented me with the Air Force Commendation Medal for meritorious service while assigned to Orlando AFB. (See citation.)

The Hump operation was the forerunner of MAT (Military Air Transport), which was the forerunner of MATS (Military Air Transport Service). MAT was formed at Stockton, California, in 1945, with operating locations at Tucson, Denver, and Seattle. I was transferred to MAT in the fall of 1945 and was in it and, later, MATS for the rest of my career.

I received a letter from General Howell M. Estes, Jr., the Commander of MATS, expressing gratitude for the years of my service and devotion to duty in the air force. (See copy.)

I was retired on October 31, 1965, with almost twenty-four years of service, twenty of them on active duty. I wanted to stay in longer, but regulations wouldn't permit it.

In spite of many close calls, I had survived three wars.

---

CITATION TO ACCOMPANY THE AWARD OF

THE AIR MEDAL
(SECOND OAK LEAF CLUSTER)

TO

ALAN W. SAUNDERS

Major Alan W. Saunders distinguished himself by meritorious achievement while participating in sustained aerial flight as a pilot in the Republic of Vietnam from 4 August 1963 to 11 December 1963. During this period, the airmanship and courage exhibited by Major Saunders in the successful accomplishment of combat support missions under extremely hazardous conditions, including continuous harassment by unfriendly forces, demonstrated his outstanding proficiency and steadfast devotion to duty. The professional ability and outstanding aerial accomplishments of Major Saunders reflect great credit upon himself and the United States Air Force.

---

*Citation for Air Medal Second Oak Leaf Cluster Vietnam.*

*Receiving two Air Medals for missions, in Vietnam, presented by Colonel Brown, Base Commander, Orlando.*

CITATION TO ACCOMPANY THE AWARD OF

THE AIR FORCE COMMENDATION MEDAL

TO

ALAN W. SAUNDERS

Major Alan W. Saunders distinguished himself by
meritorious service while assigned to the 1360th
Air Base Group, Orlando Air Force Base, Florida,
from 1 September 1964 to 31 October 1965.
During this period Major Saunders, using his
experience gained while assigned to Southeast
Asia, developed a presentation and, on his own
time, appeared before 42 civic clubs and high
school groups in Central Florida. His presen-
tations informed his audiences on USAF and
Department of Defense actions in Southeast Asia,
and promoted much pride and patriotism and
contributed outstandingly to the community
relations program of Orlando Air Force Base.
The distinctive accomplishments of Major Saunders
culminate a distinguished career in the service
of his country and reflect great credit upon
himself and the United States Air Force.

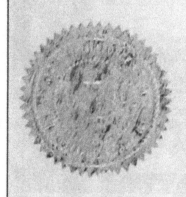

*Citation for Commendation Medal.*

HEADQUARTERS
MILITARY AIR TRANSPORT SERVICE
UNITED STATES AIR FORCE
SCOTT AIR FORCE BASE, ILLINOIS 62226

18 October 1965

Major Alan W. Saunders
1360 Air Base Group
Orlando Air Force Base, Florida

Dear Major Saunders

1. On the occasion of your retirement, I join your many friends and fellow workers in expressing gratitude for the years of conscientious service and devotion to duty that you have rendered your country.

2. The Air Force has been most fortunate in having the services of an officer possessing the professional competence that you have displayed. As you reflect on the events of your military career, you may be justly proud of your many achievements so ably performed.

3. I know that your retired status will not diminish your interest in and support of the Air Force. Best wishes for many years of happiness and success.

Sincerely

HOWELL M. ESTES, JR.
General, USAF
Commander

*Letter of gratitude from General Estes.*